AIR CONDITIONING CUTTER'S READY REFERENCE

A well illustrated, *easy to read* and *easy to apply* description of *fitting layout* that is truly a *layman's* approach to **Cutting with a System.**

The *unique* and *different* methods of presentation, which are applied throughout, literally place the *reader* at the bench, **cutting** under actual *shop* conditions.

This *book* offers the closest possible approach to simulating practical experience for which there is no real substitute.

AIR CONDITIONING CUTTER'S READY REFERENCE

Revised Edition

R. C. Morris

Member
Sheet Metal Workers
Local Union No. 108
Los Angeles, California

Business News Publishing Company
Troy, Michigan

Printed in the United States of America

Dedicated to the memory of my dad

FRANK L. MORRIS

PREFACE

This Revised Edition of the **Ready Reference** has been given a *new look* from cover to cover. However, the purpose of the *book* still remains the same: to serve, as its title implies, as a Ready Reference containing *fitting layout methods, tables, shortcuts, etc.,* appertaining to *air conditioning work* and to provide the inexperienced *cutter* with a systematic method of working that can be readily applied in *shop* practice.

This book is **not** to be looked upon as being a regular pattern drafting book since this is not the *author's* intention.

The contents of this book have been especially prepared for both the Sheet Metal Journeyman and Apprentice who aspire to become *cutters* in air conditioning work.

Due to the many requests for a more informative section on the *layout of round fittings,* the *author* has, in the course of updating and revising, added considerably more material to the Chapter titled Round Layout Development.

It is the sincere hope of the *author* that this book will in some way help to make better craftsmen of those who may read it. By doing this, the *book* will serve as a step forward towards the betterment of our *trade.*

CONTENTS

chapter I

CUTTING WITH A SYSTEM

INTRODUCTION

Although the **cutter** in *air conditioning* work is required to have the knowledge and ability to lay out and develop the more complex *fittings*, such as *transition elbows, square to rounds*, etc., it is usually the simple *fittings* and seemingly insignificant items which most often make up the greater part of the *cutter's* work day. However, their value should not be underestimated for they are just as important and as much a part of **cutting** as the most intricate *fitting* used in *air conditioning ductwork*.

Due to the simplicity of these *fittings*, plus the fact that they are most often ordered in multiple lots, they are considered *routine* work. Being routine work, they must be handled as such and developed in the shortest possible time.

No special *layout* skill is required, but the manner in which the *cutter* handles the development of simple fittings will do more to establish his status as a *cutter* than any ability he may show in developing the more complicated *fittings*.

This can be attributed to the fact that, in working up these so-called **routine fittings,** the *cutter* must apply other talents besides *fitting layout*, such as a knowledge of *shop equipment*, coordination, and ability to determine the best and most direct means of handling the work. It is here that the extent of the *cutter's* experience is most noticeable, and it may well be a deciding factor in assuring him of a steady job.

To be regarded as a *steady cutter*, he must be able to turn out a considerable amount of work each day. This does not particularly refer to working speed, but to the speed that is gained from **knowing what you are doing** and by **doing it in a systematic manner** with a **minimum loss of motion** — in other words, **cutting with a system.**

The specific purpose of *Chapter 1* is to provide the inexperienced *cutter* with *a system of cutting* that can be readily applied under shop conditions. This system helps make it possible for him to begin **cutting** at the bench without revealing his lack of experience. He

1

Cutting with a System

cannot, of course, expect to match the ability or know-how of the seasoned, experienced *cutter*. However, by adopting the *system of cutting* shown in *Chapter 1*, and applying the *layouts* shown throughout this book, he can, by adding his own ingenuity, remove practically all evidence of having **just started at the bench.**

It must be realized that the application of **cutting with a system** may have to be varied in some ways. This is due to the difference between working in a large, fully equipped shop and one of the smaller shops with a very limited amount of equipment. Lack of equipment should not affect the *cutter's* basic working procedure. If the customary piece of equipment is not available, he should select the next fastest means of doing the job. It is up to the *cutter* to apply his own ingenuity in selecting an alternate method.

A successful *cutter* is constantly thinking and planning ahead and is able to determine the correct **method of layout** to use for each *fitting* or problem.

Since the *cutter* is constantly working against **time,** he must, whenever possible, apply **short-cut methods** in making his **layouts.**

However, the *cutter* must **never sacrifice the quality of his work for quantity.** To be certain of developing and maintaining a high standard of workmanship, he should abide by the following . . .

RULE: Consider every fitting or item you make to be important enough for you to put forth your best efforts in its development, regardless of how insignificant it may seem to be.

The *fittings* that are shown and described in *Chapter 1* are relatively simple and are no problem for the experienced *cutter*. In one form or another, they are found on practically every *cutting sheet* and are handled as *routine fittings*.

Readers should note that the terms **cutting sheet, work sheet,** and **work order** all refer to the same item. While all three terms are frequently used, **cutting sheet** is almost universally accepted. These terms should not be confused with still another item called the **shear list** which is used for that purpose only.

The *work sheet* included in this section is typical of the *sheets* encountered in actual shop practice. It is from such *sheets* that the *cutter* prepares his *shear list* for the *shearman* to block out.

In preparing *Chapter 1*, no specific dimensions were used and the figures shown were selected for illustrative purposes only. Starting with a sample *cutting sheet*, the author has attempted to show the application of **cutting with a system,** in a step-by-step procedure such as that **applied in the shop.**

Layouts are made **directly on the metal,** using a *rule, scratch awl, scriber, prick punch, straightedge, trammels, dividers,* and *square.*

This *new and direct* approach to *fitting* development eliminates the need to make any transition from a *layout method* that can be readily developed on a *drawing board,* but would prove impractical for use in the *shop.* Using this new approach, the *cutter* can transfer what he reads directly onto the metal, having only to replace the measurements used in the sample problems in the book with the dimensions of the *fittings* he is required to make.

CUTTING WITH A SYSTEM

Cutting with a system begins when the *cutter* receives the cutting sheet and is applied as follows:

BASIC PROCEDURE CHART

1. Make a quick survey of cutting sheet.
2. Check gauge of metal required.
3. Establish a number for each fitting.
4. Determine the necessary stretchouts, etc.
5. Determine sizes of material in stock.
6. Determine correct methods of layout.
7. Prepare duct shear list.
8. Prepare fitting shear list.
9. Cut pieces not being sheared.
10. Notch duct and mark for fabrication.
11. Make miscellaneous items as required.
12. Layout, notch and mark fittings.

Here, in detail, are the basic procedures:

1. Make a quick survey of cutting sheet.

When the *cutter* receives the cutting sheet, he should quickly check for the following items:

    a. Are all dimensions given?

    b. If insulation or inside lining is required, are thickness and type shown?

Cutting with a System

  c. Are any special markings required, such as *zone*, *floor*, *supply*, *return air*, *etc.*?

  d. Is all necessary information shown for each *fitting*?

  e. Are *locks* or *S & D* taken off the *sheet*?

2. Check gauge of metal required.

To properly prepare the *shear list*, the cutter must first know the required gauges of the metal he will use. The gauges may be indicated on the *cutting sheet* or given to the *cutter* by the *foreman*. On very large jobs, the *foreman* may display this information within view of the cutting benches.

3. Establish a number for each fitting.

Each *fitting* should be given an identifying number. This number will identify the fitting as it moves through the layout process, fabrication and check-out before being sent to the job. At the jobsite, it will be used again to determine the location of the fitting.

Large projects may require more than one *cutting sheet* for the same job. In this event, it is better to number the *fittings* as a group and in succession, regardless of how many *sheets* the job requires.

Numbering the Fittings.

The most reliable numbering method, marked on each part of the fitting, has the *job number*, the *sheet numbers* and the *fitting number* as marked.

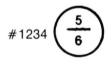

The job number is denoted by the first four digits, #1234. The top digit shown in the circle is the *fitting number* while the bottom digit is the *sheet number*.

The encircled figures identify fitting #5 from *sheet* #6. Since there is no advantage in numbering the duct, it is omitted. However, care should be taken to avoid error in counting the number of joints to be made.

Although this numbering system will identify each fitting as a complete unit, it does not keep track of the individual pieces during the *shearing* and *layout* process. Because it is so important for the cutter to know the status of work in process, a marking code will be shown at a later point in this chapter.

4. Determine the necessary stretchouts, etc.

The stretchouts required in making the *wrappers* for the *offsets*,

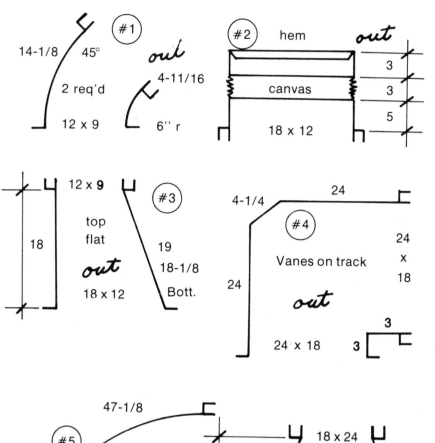

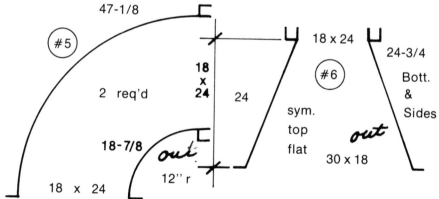

4 jts. = 30" x 18" duct — 48" lg. L. & F.
2 jts. = 18" x 12" duct — 36" lg. L. & F.
2 stubs = 12" x 9" — 16" lg. L. & F.
1 stub = 12" x 9" — 12" lg. L. & F.
4 jts. = 23" x 18" duct — 48" lg. L. & F.

Fig. 1-1

Cutting with a System

the heels and throats for elbows, and the development of the true lengths of lines, can be determined in various ways. The cutter should adopt the *method* with which he feels most confident.

The various *methods* available to the *cutter* are shown elsewhere in this book. They can be quickly found in the Index under the following subject entries:

> Math Problem No. 5 (Triangles)
> Math Problem No. 6 (Factors)
> Heel and Throat Stretchouts (Wrappers)
> Trigonometry
> Offset, determining radius point for
> Offset, stretchouts for (Wrappers)

These and other *sections* of this book should be thoroughly mastered by the *cutter* so that he can readily apply them when needed.

A sample *work sheet* is shown in Fig. 1-2. Notice that the various radius stretchouts are placed directly on or by the place required. A similar record is made of the determined true lengths which are necessary in the *layout* of *transition* or *change fittings*.

The author follows this practice for two reasons: it is easier for the cutter and it saves time in the event that any part of the *fitting* has to be remade. Often a part of a *fitting* may be spoiled in the course of fabrication or a call may come in from the *job site* for another *fitting* of the same dimensions. Marking the *work sheet* makes it easier to remake the fitting without repeating the figuring.

This practice may not be necessary for simple *fittings*. However, it can save considerable time. It also reduces the possibility of any error when making the second *layout*.

5. Determine sizes of material in stock.

Before the shear list is prepared for either duct or fittings, check the sizes of material in stock. The widths are usually 36" or 48", with the lengths being 96", 120" and 144". The quickest way is to ask the shearman. This step is particularly important when figuring out *duct*, as it usually can be made in two or three ways. In many instances, the method used is determined by the lengths of the metal sheets already in stock.

6. Determine correct methods of layout.

Take time to study the fitting before plunging headlong into its development. These few minutes are often well spent. Many *cutters* have started a fitting, then discovered, midway in the process, that they were using the wrong method of layout.

Before proceeding with *Steps 7 and 8*, these various allowances must be known.

Allowances for Seams, Edges, Locks, etc.

In preparing sample shear lists, the following allowances have been added, where required, to serve a specific purpose. These allowances are used in practically all *shops*. However, some variations may be encountered.

Before starting work in any **new shop,** the **cutter should check on these allowances** as a precautionary measure. Many *shops* have certain *equipment* which requires a slight variation to the allowances listed here. Those which change most frequently are numbers 1 and 9 of the following group:

1. Government lock or clip   1" (some shops use 7/8")
2. Flange   7/8"
3. Single Edge (duct and straight edges)   1/4"
4. All radius edges (using Easy Edger)   3/16"
5. Tapins (hand)   2-1/2" (7/8"+7/8"+3/4")
6. Tapins (machine)   1-3/4"
7. Caps   7/8"+7/8" (add 3-3/8"+ req'd size: allows 1/8")
8. Pittsburg (hand)   1-1/4"
9. Pittsburg (machine)   1" (some shops use 15/16" or 1-1/16")
10. Patch door   1"+1"+5/8" (add 5-1/4" to size of req'd door)
11. Snap lock (20 ga.)   1-5/16"
12. Snap lock (single)   1/2"
13. Snap lock (24 ga.)   1-1/16"
14. Snap lock (single)   7/16"

On *cutting sheets* where duct is included, the *cutter* should make a separate *shear list* for the *duct* (see **Duct Layout**).

Certain dimensions are sometimes shown as $\pm 1/2$". This indicates that the dimension is not critical and can vary either way by the amount of the fraction. This allowance gives the *shearman* a chance to use up all possible scrap material. Often, a piece of metal may be within a quarter of an inch of the given size. The *shearman*

Cutting with a System

will not be able to use this undersize piece of scrap unless some indication is given that this particular measurement is not critical.

When **duct length** is given as 48" or 36", it is assumed that the **net** or **finished length** will fall short by the allowance for the **lock** and **flange,** or **S. & D.** The actual lengths are closer to 46-1/4" and 34-1/4", depending on the width of the sheet which usually runs about 1/8" to 1/4" oversize.

When the finished size must hold to the 48" or 36" length, as sometimes happens, this is usually specified.

7. Prepare duct shear list.

Duct requirements are listed at the bottom of *work sheet #1,* Fig. 1-1. The first is . . .

4 = 30" x 18" x 48" long

Although this 30" x 18" *duct* can be made in one piece by using 120" metal, the length makes it difficult to handle. Making it up in half-sections, or three-sided with a drop-in panel, is an easier method of fabrication.

The *author* prefers a three-sided *duct* that adds up as follows: 18" + 18" + 30" to which the *pittsburg* lock allowance for both sides must be added. Assuming the *pittsburg* to be 1" on each side, this makes the total *shear* size . . .

68" x 48" = 4 pcs. req'd.

The drop-in panel will be 30-1/2" wide.

30-1/2" x 48" = 4 pcs. req'd.
2 = 18" x 12" x 36" long.
        (net length under 36")

This 18" x 12" *duct* can satisfactorily be in one wrap-around, which is 60" + 1" + 1/4" or a total of . . .

61-1/4" x 36" = 2 pcs. req'd.

2 = 12" x 9" x 16"-long stubs
1 = 12" x 9" x 12"-long stubs

These are easily made in one piece, with the total girth or material required being 43-1/4".

43-1/4" x 17-7/8" = 2 pcs. req'd.
43-1/4" x 13-7/8" = 1 pc. req'd.

4 = 24" x 18" x 48" long

This duct can be made in one piece but is easier to handle when made in either half joints or as three-sided *duct* with a drop-in panel. In this instance it is made in half joints which require a length of 43-1/4".

43-1/4" x 48" = 8 pcs. req'd.

8

Cutting with a System

The completed *duct shear list* is shown here as it should be prepared by the *cutter*. **As a precaution, the list should be rechecked** before turning it over to the *shearman*.

Job #1234	Sheet #1		all w/L. & F.
4 = 68 x 48	30 x 18		24 ga.
4 = 30-1/2 x 48	30 x 18 		
2 = 61-1/4 x 36	18 x 12		26 ga.
2 = 43-1/4 x 17-7/8	12 x 9 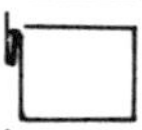		
1 = 43-1/4 x 13-7/8	12 x 9 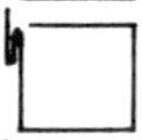		
8 = 43-1/4 x 48	24 x 18	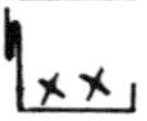	24 ga.

Fig. 1-2

NOTE: The symbols for the various duct shapes need only be roughly drawn, as shown. The shearman transfers these markings from the shear list to the sheared metal for the notching and forming process (Fig. 1-2).

Before proceeding with **Step 8**, the preparation of the *fitting shear list*, the cutter should study the many advantages of the **Marking Code** and be capable of applying it as a normal *shop* routine.

Purpose of the Marking Code

To avoid cutting too many or too few pieces, the *cutter* must have some means of knowing the exact location of each piece at all times.

The *reader* may consider this a trivial matter but it is **necessary and important** to the cutter, especially when the *shop* is extremely busy and many different *jobs* are in process at the same time. The cutter may have to work on several jobs at the same time, leaving some of them unfinished with the material set aside.

When there are several *cutters* in a *shop*, the *shearman* must keep each one supplied with the material shown on his *shear list*. To keep the material flowing, he will only cut a part of each man's *shear list*, just enough to keep every *cutter* busy. Here is another

Cutting with a System

reason why it is so important for the *cutter* to keep an accurate check on all material he orders cut at the *shear* and on each individual piece of a *fitting* as it is completed.

Another fact to consider is that the cutter will seldom be able to complete each *fitting* in turn, since his material will not be sheared in the order in which his *shear list* was written.

In cutting the contents of the *shear list*, the *shearman* first uses the scrap or drop-offs he has on hand. This means, of course, that the material which is delivered to the *cutter's* bench may have the sheared pieces stacked in random order so that no one fitting can be completed without digging through the stack. This is considered unnecessary handling of the metal, to no real advantage. Therefore, the *cutter* works the pieces in the order in which they are stacked.

Unless it is a rather complicated *fitting*, the *cutter* will throw the pieces onto the table as they are laid out, ready for the *brake*. Keeping track of each piece under these conditions is obviously quite a problem for the *cutter*.

It was many years ago that the *author* set up his own **Marking Code** to handle this situation. The **Marking Code** shown here is a simplified method for maintaining a constant, accurate check, directly on the sketch of the *fitting* shown on the *cutting sheet*.

By using this *code*, the cutter knows exactly where he stands at all times. Even if the *cutter* is pulled off the *job*, he can come back to it at any time and continue from where he left off.

This *method* is practically foolproof if the *cutter* abides by one definite . . .

RULE: Do not mark any part of the fitting as being finished until it has been placed on the table, ready for the brakeman.

This also means that, even though one or more parts of a *fitting* may be finished and laid aside, **these parts should not be marked off the cutting sheet,** until the complete fitting has been cut.

Violating this rule is about the easiest way to become mixed-up, especially **if the cutter is taken off the job** for any length of time. However, when **holding steadfastly to the rule,** there will be very little chance of error.

When preparing the *shear list*, mark each fitting in the following manner:

MARKING CODE

oul Placing the word *out* directly on or at the *fitting*, **without crossing** the letter *T*, indicates that all of the fitting has has not been taken off the *cutting sheet*.

After marking each piece on the *shear list*, the cutter can refer to this list at any time to determine which pieces are not being cut.

out Placing the word *OUT* directly on or at the *fitting*, and crossing the letter *T*, indicates that the entire *fitting* has been taken off the *cutting sheet*.

This is all that is required until the sheared material has been brought to the *cutter* and he is ready to complete the *layout* of the various pieces.

The word *OUT*, previously placed on each *fitting*, is used again with various horizontal and vertical lines. These lines indicate that certain pieces of the fitting are finished and on the table ready for the *brakeman*.

out̄ Placing a line **above** the word indicates that the **top** of the *fitting, elbow cheek, transition, etc.*, is finished.

out̲ Placing a line **below** the word indicates that the **bottom** of the *fitting, elbow cheek, transition, etc.*, is finished.

out| Placing a line at the **right side** indicates that the **throat** of the *elbow*, or **right side** of the *transition*, is finished.

|out Placing a line at the **left side** indicates that the **heel** of the *elbow*, or **left side** of the *transition*, is finished.

Start by indicating the first part of the *fitting* being cut. As each additional piece is cut, add the appropriate **mark** until all of the four pieces are cut.

|out| Lines on all four sides of the word *OUT* means that the *fitting* is completed and on the table **ready for fabrication.**

When duplicates of the same *fitting* are required, a small number is placed along side each line to indicate the number of pieces completed.

This simple **marking code** enables the cutter, at any time, by

Cutting with a System

glancing at his **cutting sheet,** to determine which parts of each *fitting* he has completed and given to the *brakeman* and also how many.

8. Prepare fitting shear list.

Here is the *shear list* as it should be prepared for the *fittings* on **work sheet #1**, Fig. 1-1. Notations on the *shear list* indicate where each piece belongs, making it easy to pick out the pieces as required.

Shear list	Job #1234	Sheet #1
2 = 16 x 11		heel #1 26 ga.
2 = 6-9/16 x 11		thro. #1
1 = 60-3/4 x 4-3/8		canv. conn. #2 24 ga.
1 = 60-3/4 x 6-7/8		canv. conn. #2
1 = 30-1/2 x 19-7/8		top & side #3
1 = 20-7/8 x 14		w/2 pitts. #3 lg. sd.
1 = 20-1/8 x 19-1/4		w/1 pitts. #3 bott.
1 = 54-1/8 x 20		heel sq. ell #4
1 = 7-7/8 x 20		thro. sq. ell #4
2 = 28-1/4 x 28-1/8		chks. sq. ell #4
2 = 49 x 26		heel #5
2 = 26 x 20-3/4		thro. #5
4 = 32 ±1/2 x 32 ±1/2		cheek #5
1 = 25-7/8 x 30-1/2		flat top #6
1 = 26-5/8 x 30-1/2		bott. #6
2 = 26-5/8 x 26-1/2		sides #6

Fig. 1-3

NOTE: Development of the Shear List has been included in Step 12, under the caption, Development of Work Sheet #1.

9. Cut pieces not being sheared

Cut the radius cheeks, and any other irregular pieces while waiting for material to be returned from the shear.

10. Notch duct and mark for fabrication.

Duct should be notched either by **hand** or at the **duct notcher**, according to standard *shop* practice. This is usually done while material for *fittings* is being cut.

11. Make miscellaneous items as required.

Include such miscellaneous items as *dampers, doors, angle iron frames,* if required, etc. These can be made after **Step 12** is completed, providing the material has been cut for the *fitting.*

No definite **rules** govern the order in which to work. Priorities must be established by the *cutter* in response to the conditions which exist at the time.

12. Layout, notch and mark fittings.

The remainder of this chapter is a detailed description of how the fittings in Fig. 1-1 are developed from the *shear list*.

Development of Work Sheet #1

Fitting #1   **45° Elbow, 2 required**

The cheeks for this *fitting* are cut at the bench from scrap metal, usually while waiting for the other material to be cut at the *shear*. It is common practice to cut several pieces at one time, following the *layout* made on the top piece.

If a *bandsaw* is available, all four pieces can be clamped together with the special clamps that are designed for the saw. If the *bandsaw* is already in use, the cheeks can be cut at the bench. Clamp two pieces together with the vise grips and cut with the *Unishear*. Then repeat the process with the other two pieces.

The *bandsaw* should be used whenever there are a large number of pieces to cut. Depending upon the capacity, the average *saw* can handle a 1/2" high stack of pieces at one time.

It is seldom practical to bother with the saw when cutting just a couple of pieces. These can be added to the shear list since it is a simple matter to estimate the blockout size.

The length is determined by adding the radius to the width of the cheek. To determine the height, refer to the *trigonometry chart*. Find the sine function for 45° which gives the figure .7071. This figure is multiplied by the total of the width of the cheek and the radius; the result being 12.7278 or approximately 12-3/4". After allowances for both lock and flange, this piece blocks out at about 14-1/2". This application of trigonometry is described in several places throughout this book and the reader should make certain he understands it thoroughly.

Fig. 1-4 shows the blockout for the cheek of *fitting #1*. It is made up as follows: use a 1" *scriber* across the bottom of the longest width, to allow for the lock. Scribe in about 1/2" from the right-hand side (or left if preferred). The intersection *a* of these two is the *apex* for development of the cheek. From point *a*, measure out 9" for the throat radius *b*, then 12" more for the heel radius *c*.

Cutting with a System

There are several methods for determining the 45° angle. Since this is a small fitting, a protractor or any other method is adequate. After establishing the 45° line, allow another 7/8" for the flange. Use trammels to draw in the arcs of heel and throat. Then scribe in the cutting lines which are 3/16" outside of the original arcs.

Mark the sheet, as shown, and include the job number, #1234, which was omitted here due to a lack of space. The shaded area is to be cut away and the corners notched. Each of the fittings

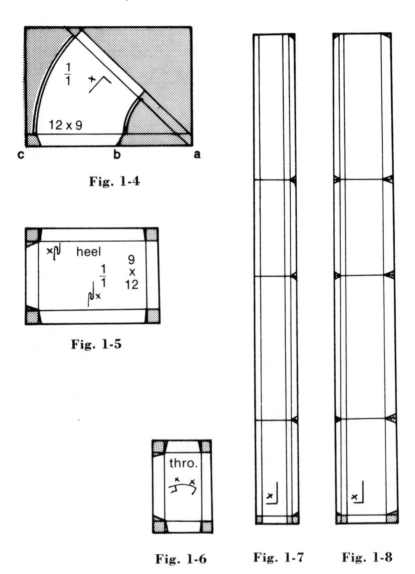

Fig. 1-4

Fig. 1-5

Fig. 1-6   Fig. 1-7   Fig. 1-8

requires a *left* as well as a *right* cheek which is obtained by turning the right cheek over and marking the other side.

Notice the word *OUT* as shown at *fitting #1* on the work sheet. The letter *T* was purposely left uncrossed to indicate that not all of the *fitting* was ordered at the *shear*.

Fig. 1-5 shows the heel *layout* which requires a 16" x 11" piece of metal. The *stretchout* is 14-1/8" + 1" + 7/8" = 16". The width, 9" + 2" = 11", being the wrapper which includes the *pittsburg*.

Use a 1" *scriber* off the 3 sides to mark the *pittsburg* and lock edges. A 7/8" *scriber*, used off the opposite end, allows for the flange.

Notch out the corners, as shown by the shaded area, then mark as required to complete the heel.

Fig. 1-6 shows the throat *layout* which is the same as the heel.

Due to a lack of space, the job *number* has been omitted from these figures. In actual practice, always place the *job number* on each piece.

Fitting #2   **Canvas Connection**

This *fitting* is shown in the *shear list* and again, in block-out form, in Figs. 1-7 and 1-8. To describe the *layout* procedure now would be a repetition of the *layout* given in the *Canvas Connection Section* (see **Index**).

The connection size is 18" x 12". Therefore, the stretchout is
18" + 12" = 30" x 2" = 60" + 3/4" = 60-3/4"
The *spotwelding* allowance need not be exact and 5/8" or 3/4" is a good width for lap.

The *shear list* shows the length to be 60-3/4" with a width of 4-3/8" for one side and 6-7/8" for the other.

The work sheet, in Fig. 1-1, specifies both a 5" and a 3" width.
3" + 1/2" (for *hem*) + 7/8" = 4-3/8"
5" + 1" (for *gov. lock*) + 7/8" = 6-7/8"

Fitting #3   **Transition**

Fig. 1-9 shows the complete *layout* for the top and one side. The shaded area is to be cut away. The *shear list* gives this section as . . .

30-1/2" x 19-7/8".

Cutting with a System

The largest figures determine the other dimension, plus 1/2" for the single edges. In this case 18" + 12" + 1/2" = 30-1/2".

Scribe a 1" line across the top for the lock and a 7/8" line across the bottom for the flange. Use a 1/4" scriber across the flange line at each end; then measure off 18" from the left and 12" from the right, being certain to have the 18" width on the left-hand side.

From this point, scribe a perpendicular line to the top of the sheet. Then, measure off 12" to the left and 9" to the right of this line. This will establish the four outside points which are connected, as shown in Fig. 1-9. Holding a *bench rule* or *straightedge* against the points established for each side, connect with a straight line.

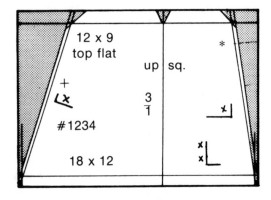

Fig. 1-9

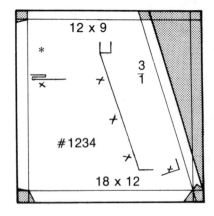

Fig. 1-10

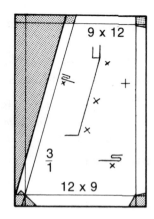

Fig. 1-11

Place the *straightedge* on the inside of the *fitting*. Both the connecting line and the outer cutting line can be made in one pass with a special 1/4" *scriber* made for scribing across the sheet (see **Index** for *scribers*.)

Shaded areas are cut away and notched, and the flange, center line, and the 1/4" single edges are prick marked. Mark as required to complete this part of the *transition fitting*.

NOTE: The plus sign and the asterisk, any other marks may be used, are placed on the left and right edges of the piece shown in Fig. 1-9. Matching symbols are also placed on the left and right edges of the mating pieces shown in Figs. 1-10 and 1-11. This code makes it much easier to sort out, fabricate and match pieces of fittings with slight variations in size or rise.

When laying out *transition fittings*, such as these, there is always less chance of error if the *cutter* lays out each fitting on the metal as it appears on the *work sheet*. If this had been the case with this *fitting*, the *layout* would be on the outside with the marks being placed on the other side.

Fig. 1-10 shows the bottom piece of the *transition fitting*. The block-out size is shown on the shear list as . . .

20-1/8" x 19-1/4."

This piece has the *pittsburg* on one side and the 1/4" single edge on the other.

The bottom rise is 3" in 18", making 18-1/4" the true length. With 1-7/8" for the lock and flange, this adds up to 20-1/8". The width equals 18" + 1" + 1/4" a total of 19-1/4".

Scribe a 7/8" line across the bottom for the flange and a 1" line up the left-hand side for the *pittsburg*. Scribe another 1" across the top for the lock. From the left 1" line, measure off 18" across the bottom, 12" across the top.

Place the *bench rule* against the two points and connect them in the same manner that was applied to the top section of this *fitting*.

The shaded area is to be cut away, then marked as shown in Fig. 1-10.

This being the bottom, which has a 3" rise up, the lock end should be given a slight kink down, while the flange end will be slightly under square.

Prick mark the flange, 1/4" edge and lock end.

Cutting with a System

Fig. 1-11 shows the right or long side of the *transition* which the *shear list* shows to be . . .

20-7/8" x 14", with 2 pittsburgs.

Referring to the *work sheet*, the true length is 19" + 1" + 7/8" = 20-7/8". The width requires 12" + 2" for the *pittsburgs*.

From the *pittsburg* edge, measure over 9" at the top and 12" at the bottom. Using a *bench rule* and *scriber* in the usual manner, connect these two points. The 1" *scriber* is required for the *pittsburg* lock edge.

The shaded area is to be cut away and the marks placed as shown in Fig. 1-11. Prick mark both the flange for 90° and lock, which requires a slight kink in the *brake*. This completes *fitting #3*.

Fitting #4   **Square Elbow**

Fig. 1-12 shows the *layout* for the cheek. Two are required, a right and left. This is accomplished by making two alike, then turning one over to be marked on the other side.

Referring to the *shear list*, the block-out size for the cheeks is . . .

28-1/4" x 28-1/8".

The 28-1/8" width is figured as follows: 24" + 3" + 7/8" + 1/4"= 28-1/8".

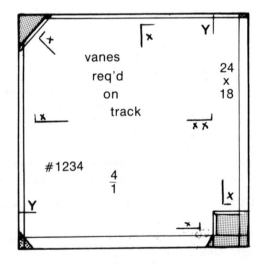

Fig. 1-12

This overall width allows for a 7/8" flange. The 28-1/4" width given for the other overall dimension allows 1" for the lock.

Scribe a 1" line across the base and, on the right-hand side, a 7/8" line for the flange. Also scribe a 1/4" line across the top and left side for the single edges.

The cut-off corner is 3" and 3" off the 1/4" lines, at the upper left-hand corner, Fig. 1-12. This makes the 45° line 4-1/4" across. Many shops use a 4" and 4" corner which makes the diagonal line 5-5/8" long.

NOTE: A template or pattern for this corner is shown elsewhere in this book, listed under Time Saver Patterns.

The indicated prick marks, shown at Y, are for the *brakeman's* use when he makes up the 1/4" edges of the throat by malleting them over the edge of the *brake*. Some shops have machines to turn these edges. Then, these marks are not necessary.

Prick mark the flange and 1/4" single edges. Then add the marks as shown in Fig. 1-12. Since a right and left cheek are required, make 2 cheeks, both alike, being certain to turn one over and place the marks on the other side. Vanes are required on track. This subject will be discussed later in another *Chapter* of this book (see **Index**).

Fig. 1-13 shows the throat, which is shown on the shear list as . . . 7-7/8" x 20".

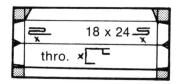

Fig. 1-13

The first dimension is 3" + 3" + 1" + 7/8", a total of 7-7/8". The width of the elbow is 18", plus 2" for the *pittsburgs*, a total of 20".

Scribe a 1" line across three edges, with the longest edge being for the *lock* and the two short sides for the *pittsburgs*. The bottom 7/8" edge will be for the flange.

Measure off 3" from the 1" lock line to divide the center for the throat. Extend this mark up from the bottom, then prick mark the flange and center break.

Cutting with a System

The shaded area shows the notching required. Place the various marks as shown, including the *job number*, to complete the throat.

Fig. 1-14 shows the heel for the elbow which the shear list shows as . . .

54-1/8" x 20".

This is figured as follows: 1" + 24" + 24" + 4-1/4" + 7/8" = 54-1/8" overall.

Scribe a 1" line across the top and bottom, or long sides, for the *pittsburgs* and another along the left side for a lock. Scribe a 7/8" line across the right side for the flange.

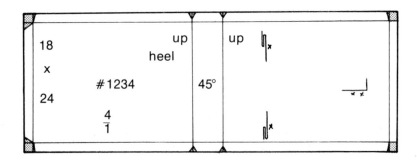

Fig. 1-14

Measure in 24" from both of the lines, which should leave 4-1/4" through the center.

The shaded area shows the notches to be cut out. Prick mark the flange and the two center 45° breaks. To complete the heel *layout*, mark as shown in Fig. 1-14.

Fitting #5   **Radius Elbow**

Fig. 1-15 shows the cheek for this *elbow*. As a rule, radius cheeks are usually cut at the bench from a full sheet of metal. The *cutter* can save material by nesting the pattern which is the first piece cut out. However, on occasions when the fitting is required in a hurry, considerable time can be saved by blocking out the cheeks as shown. Referring to the shear list, it shows . . .

32" ±1/2" x 32" ±1/2"

as the blocked out piece for the cheek. Since the metal is cut over-size, the *fitting* can be laid out from either edge.

20

Scribe a 1" line across the bottom for the lock and a 7/8" line off the right side. From the intersecting point of these two lines, measure off the given 12" radius. From this new point, measure off the 18" width for the cheek. Using the *dividers* or *trammel points*, scribe the quarter circles which are the heel and throat. Measure off 3/16" on the outside of the heel and on the inside of the throat. It is not necessary to measure off the 3/16" since this dimension can be fairly closely judged, without a rule. From these points, scribe the lines for the single edges which are also the cutting lines of the pattern.

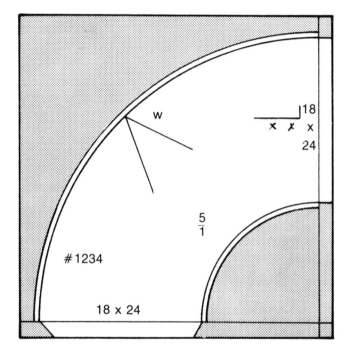

Fig. 1-15

These edges are the allowance for the **easy edger.** Prick mark the approximate center of the cheek at the heel, as shown at *w* in Fig. 1-15, as the location point in panelling the cheek; also prick mark the flange.

The shaded area shows the material to be cut away. Mark the fitting number and other marks, as required, to complete the cheek. The opposite cheek is made in the same manner as described for the square elbow.

Cutting with a System

Fig. 1-16 shows the heel for the elbow, which the shear list shows to be . . .

49" x 26".

The stretchout should be 47-1/8" + 1" + 7/8" = 49". The width is 24" + 2" for the *pittsburgs*, or 26".

The heel stretchout can be taken from the *Chart* in this book which lists *Stretchouts for 90° Arcs* or off the 4 ft. bench rule. Another method can be found under *Math Problems* (see **Index**).

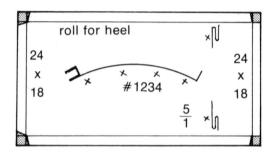
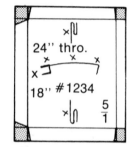

Fig. 1-16   **Fig. 1-17**

Scribe a 1" line across the bottom and top for the *pittsburgs*, and a 1" line down the left side for the lock. Scribe a 7/8" line on the right side for the flange. Cut out corner notches in the usual manner and prick mark the flange.

Mark as required, showing piece to be rolled for the heel.

Fig. 1-17 shows the throat for the elbow which is given on the shear list as . . .

20-3/4 x 26".

This was calculated in the same manner as the heel, starting with a throat stretchout of 18-7/8".

Scribe a 1" line across the bottom and top for the *pittsburg*, and another on the left side for the lock. On the right side, scribe a 7/8" line for the flange and prick mark.

Cut out the notches as shown and mark the piece to be rolled for the throat, to complete the throat *layout*.

Fig. 1-18 shows an enlarged corner of the lock and *pittsburg*, with the shaded area being properly notched for a government lock.

22

Cutting with a System

Fig. 1-19 shows an enlarged corner of the flange and *pittsburg*, with the shaded area indicating how the notch should be cut for a flanged end. Notice in Figs. 1-18 and 1-19 that the angle of the cut across the *pittsburg* is only about 1/8" at the outer edge.

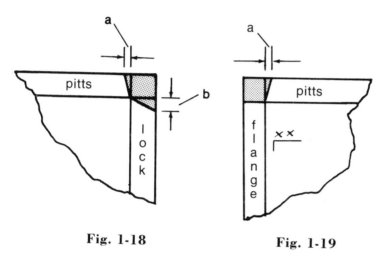

Fig. 1-18   **Fig. 1-19**

This 1/8" is nominal and need not be measured. The cutter usually makes the cut by sighting an approximate width. The same holds true with the cutback at the lock allowance edge shown in Fig. 1-18. This is usually cut back about 3/8" to 1/2", being variable to the same extent.

The flange end is usually cut off square and the lock end is angled off, so that the lock can be driven down all the way. This also is a means of quickly distinguishing the lock end from the flange, during the course of fabrication.

Fitting #6   **Transition**

Fig. 1-20 shows the **flat top** of the *transition*. This piece could either be the top or bottom as the *fitting* is symmetrical. The shear list shows a required block out size of . . .
    25-7/8" x 30-1/2".

The **Work Sheet** shows that *fitting #6* has a given height through the center of 24"; with a lock and flange, 24" + 1" + 7/8" = 25-7/8".

The width is 30" + 1/4" + 1/4" = 30-1/2" which allows for the single edge on each side.

Scribe a 7/8" edge across the bottom for the flange and a 1" line across the top for the lock.

Cutting with a System

Working off the outside edges of the right and left sides of the metal, use a 1/4" scriber and place a short line across the bottom flange line. This establishes the 30" width at the flange.

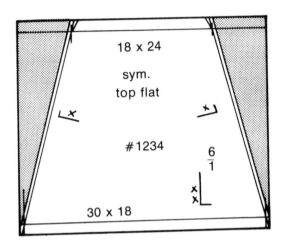

18 x 24

sym.

top flat

#1234

$\frac{6}{1}$

30 x 18

Fig. 1-20

Locate the center of the flange line and, using a *square* off the bottom, establish the center point on the lock line. Measure out 9" each way, making the required 18".

NOTE: *This method of measuring from the center line assures that the top and bottom of the fitting will be in an exact relationship. When taking the top measurement from either side, there is always the possibility that the metal has been sheared out of square. As a result, the fitting will be out of square.*

After all the outside corners are located, the sides are connected by placing the *bench rule* at the top points and scribing in the usual manner. Use the special 1/4" *scriber* across the sheet for the single edge, repeating the same process on the other side.

The shaded area is to be cut away as shown. Prick mark the flange and single edges. Mark as shown to complete the bottom *layout*.

Fig. 1-21 shows the bottom of the transition which was given on the shear list as . . .
    26-5/8" x 30-1/2".
The layout method for this piece is the same as that used for Fig. 1-20. A mere 3/4" of added length is the only difference.

Prick mark all edges, as the lock edge also should have a kink down because of the 6" rise.

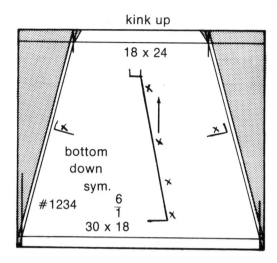

Fig. 1-21

Fig. 1-22 shows the left side of the transition which should be . . . 26-5/8" x 26".

Scribe a line 7/8" off the bottom edge for the flange. Scribe 1" lines for the lock on top and the *pittsburg* on both sides. This makes the *pittsburg* on the left side or flat top part of the fitting.

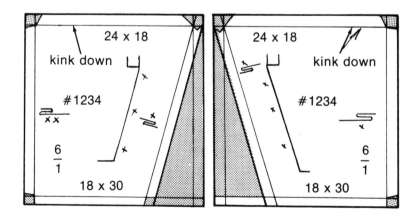

Fig. 1-22   **Fig. 1-23**

Cutting with a System

At the top of the sheet, the lock end, measure off 24". On the bottom, measure off 18", both measurements being off the same 1" line. Use the *bench rule* and the 1" *scriber* for across the sheet to scribe the *pittsburg* in the usual manner.

Cut away the shaded area, as usual. Prick mark the flange edge and the lock edge which requires a slight kink down. Mark as shown.

Fig. 1-23 shows the right side of the *transition* which is identical in all respects except for the marking which must be placed on the opposite side. Using the left side as a pattern makes it unnecessary to *lay out* the right side.

SUMMARY

This Chapter has had a twofold purpose: to provide the reader with the necessary knowledge and know-how, at the same time to give him the confidence needed to apply all of his newfound ability in actual *shop* practice.

The cutter will, at various times, be starting to work in *shops* that are **new** to him, which means that the **shop methods may differ** from those to which he is accustomed. Although the cutter may have to rearrange his **working procedure** this should not necessarily cause him to deviate from his **system of cutting.**

Any attempt to change his **system of cutting** would defeat the purpose of this chapter.

chapter II

STRAIGHT TRANSITIONS
OR CHANGE FITTINGS

INTRODUCTION

One of the most widely used *fittings* in *air conditioning ductwork systems* is the **straight transition** or **change fitting**. As its name implies, the basic purpose of this *fitting* is to make a change in **duct size.** In the process of making this transition from one size to another, this same *fitting* can be used to offset the *line* of *duct* either **up, down** or **sideways**.

As a result, this *fitting*, in its countless variations of **design** and **sizes**, is made more often than any other *fitting*. Although this is one of the easiest *fittings* to develop, it is also one of the easiest *fittings* for the *cutter* to unwittingly make some simple mistake, that usually goes unnoticed until it is being fabricated. This is especially true with very large *fittings*, where some of the pieces have to be made in several parts, or when the **rise** or **drop** of the *fitting* is extremely sharp, possibly offsetting at the same time. In these cases, extreme caution must be taken in developing the *layout*.

Most mistakes are due to carelessness, **not** the *cutter's* lack of ability. Therefore, it is wise to hold each piece of the *fitting* aside, until all pieces are completed, so that the corner **miters** and **markings** can be quickly rechecked.

Although it is **not** a required practice or a necessary part of the *layout* being developed, the *author* makes a rough freehand *sketch* of the *fitting* in *side elevation*, as shown in Fig. 2-2. The various dimensions are noted in their respective places on this *sketch*, along with the *true lengths* of the sloping *sides*.

While serving as a reference, the *sketch* also aids the *cutter* in visualizing the actual **rise** and **offset** of each separate piece of the *fitting* in its relation to each other. When preparing the **shear list**, notes can be made on the *sketch* regarding how each piece is planned on being used, plus any other small item which might be overlooked later on. The few minutes required to make this *sketch* on a scrap of paper will minimize mistakes and reduce *layout* time.

Transitions

There are actually three methods for determining the **true lengths** of **lines**: by using *mathematics*, by measuring them out on the metal, or by means of a **chart**. The *chart* is the most popular method with *cutters*, and is also preferred by the *author*.

For those who prefer using *mathematics* for this procedure, a Mathematics Handbook, containing a **Table** of **Squares** of **Mixed Numbers**, is a **must**. Without it, this method is too lengthy and slow, making it impractical for *shop* use.

The use of *mathematics* has been applied to the first *transition*, referred to as Example #1, so that its application will be understood. However, for the remainder of this *Chapter*, the **true lengths** will be taken from a *Chart*.

The various **transition fittings** which make up this *Chapter* are **not** laid out to any given *scale*, but the illustrations and dimensions have been proportioned as considered best for explanatory purposes. However, the various calculations applied are correct. The *transition fittings* in this *Chapter* have been referred to as Example #1, #2, #3, etc., for reference purposes only. These numbers only indicate an example or another way in which this type of *fitting* might appear on a *work sheet*, and how it should be developed. No attempt should be made to try to remember the *layout* described as being any specifically numbered method to apply in actual *shop* practice. The theory and method of application are both correct, therefore, it is important that the **method applied** to each **example** is thoroughly understood, to the extent of recognizing **where** and **how** to apply it.

TRIANGULATION

Since **triangulation** is the **method** of **layout** most often used in *air conditioning* work, the *cutter* should be able to apply this *method* to any *fitting*. That is, he must be able to recognize **when** and **where** *triangulation* can be **applied**, and not have to limit its use to only those *fittings* with which he is familiar.

Triangulation is a *method* of *layout* for developing *fittings* which is as individual as **parallel** and **radial line** development. Each *method* is, in itself, distinctly different from the other and, likewise, is **applied** in a different manner, each being adapted to different types of *fittings*.

This section of Chapter II is for the benefit of those who may not fully understand the *triangulation method of layout*, how it is applied and the kind of *fittings* it is applied to. It is important to the *cutter* that he knows the reason for what he does rather than just following the given instructions for developing a *fitting*.

Triangulation is, as it name implies, based upon the development, by means of **triangles**, of irregularly shaped *fittings* that have either curved or plane surfaces. The actual *layout* procedure is as follows:

Using the given dimensions and outline as shown, make a *layout* of the *fitting* to be developed. Then divide the surface area into *triangles*. Since this drawing is made to the given dimensions, it represents the **flat side**. Therefore, the **lines** which divide the **layout** surface area into *triangles* are the **base** of the *triangle*. The **height** or **rise** which occurs within the length of each *line* determines the length of the *triangulated line*, also referred to as the **hypotenuse**.

Using the new lengths of the *triangulated lines*, reconstruct the *triangles* in the same manner, following in progressive order, according to their position in the **flat pattern**.

The word *triangulate* means not only to divide into *triangles*, but also to make *triangular*. Therefore, since it is necessary to form a *triangle* in order to find the **true** length of a **line**, this process is known as *triangulating* a **line**. To apply the *triangulation method*, it is necessary to *triangulate* various **lines** and **sides** during the *layout* process.

By referring to the *triangulating* of both **lines** and **sides**, the *reader* should find it easier to associate the different triangulated *fittings* and thus remember their *layout methods*. Some of the best examples are: the **lines** of a flaring *roof jack*, the **lines** of a *transition elbow*, and the **sides** of the *straight transition fitting*. Any other *fittings* that are encountered will be similar to one of these three *fittings*, as far as their development is concerned.

Each of these *fittings* represents a distinct difference in the development of the **true length lines**. However, the **roof jack** and **elbow** are quite similar in one respect: the various **lines** are *triangulated*. On the other hand, the **straight transition** has the **sides** *triangulated*, with the end measurement laid off from a given point. In the *straight transition* the **sides** are the *hypotenuse* of the *right triangle*, making this a reverse procedure.

Although the *triangulation* of these *fittings* is applied to the surface of the **sides**, the same principle is used on the *transition*

Transitions

elbow. However, in this case, the **lines** or point-to-point *stretchouts* are *triangulated.* The same applies when *laying out* the *square to round* and similar type *fittings.*

It is easy to see how the relationship is established between the sides of the *triangle.* Knowing only two dimensions of the *triangle,* it is possible to determine the unknown length of the third side. The method is shown in *Math Problem No. 5* (see **Index**) with a rule being given for determining each side.

STRAIGHT TRANSITION
Example #1

The **transition fitting** referred to as Example #1 is shown in Fig. 2-1 as it might appear on a *work sheet.* This example is typical of the *transition fitting* having all four sides offsetting, with each side being different.

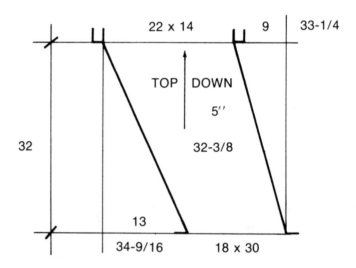

22 x 14   9   33-1/4

TOP | DOWN

5''

32   32-3/8

13

34-9/16   18 x 30

Fig. 2-1

Fig. 2-2 is the freehand *sketch* which was discussed in the *introduction* to this *Chapter.* As this is the first *fitting* showing this *sketch*, attention is called to the advantages that were described. For this particular *transition fitting* which has all sides different, it is especially helpful.

Whenever the *fitting* has all sides offsetting, it is best to select the sides having the least amount of offset to place the *pittsburg*

Example #1

edges. Referring to Fig. 2-1 and Fig. 2-2, note that the sides are best for the *pittsburg* on this *fitting*.

When preparing this or any other *straight transition fitting* for the *shear list*, all lengths should be exact and include the allowances for the **lock, flange, pittsburg,** and **single edge,** as required. This makes it easier to work up the *fittings*, since it is possible to use **scribers** off the edges of the blocked-out material, thereby reducing the **layout time** considerably.

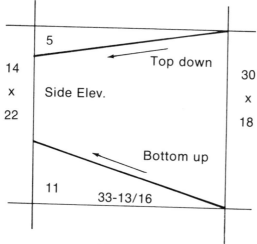

Fig. 2-2

In preparing the **shear list,** the various allowances have been figured as follows: **pittsburg** = 1″, **flange** = 7/8″, **single edge** = 1/4″. These same allowances have been used throughout the *book,* whenever required.

The sides, the top and bottom of this *fitting* are *triangulated* in the regular manner. The unknown side of the four *right triangles* which are developed from this *fitting* is found by using the procedure shown in *Math Problem No. 5.* Here is the *mathematical method* for determining the *true length* or *hypotenuse,* the part of the *right triangle* that represents the side of the *transition fitting.*

RULE: Add the square of the offset to the square of the fitting height. From this sum, extract the square root to determine the true length of the side.

Applying this **rule** to the *fitting* shown in Figs. 2-1 and 2-2, we use the following dimensions:

Transitions

Top	Down — 5″		$5 \times 5 = 25$
	Fitting height — 32″	(from **Table**)	$32 \times 32 = 1024$
			sum 1049

The sum of the two square totals 1049. Refer to the **Table** and find the closest number to 1049. This proves to be 1048.14. The square root of 1048.14 is 32-3/8″, which is the true length of the line.

Applying the same method, the other three sides or angles produce these results:

Short side

$$9 \times 9 = 81$$
$$32 \times 32 = 1024$$
$$1105$$

The **Table** shows that 1105.56 is the closest figure, and the true length proves to be 33-1/4″.

Bottom

$$11 \times 11 = 121$$
$$32 \times 32 = 1024$$
$$1145$$

Again the **Table** shows the closest number to be 1145.40. The true length is then 33-27/32″.

However, a workable figure must be selected. To simplify the fraction, reduce it to 26/32″ or 13/16″, which makes the *bottom* length 33-13/16″.

Long side

$$13 \times 13 = 169$$
$$32 \times 32 = 1024$$
$$1193$$

Again the **Table** shows the closest number is 1194.57, with the true length 34-9/16″.

Referring back to Figs. 2-1 and 2-2, the block-out sizes can now be prepared for the *shear list*. Since the *top* and *bottom* have the *single edge* and the *pittsburg* is on the *sides*, the *shear list* reads as follows:

Top	34-1/4″ x 31-1/2″
Bottom	35-11/16″ x 31-1/2″
Short side	35-1/8″ x 32″
Long side	36-7/16″ x 32″

As a reminder, all *fittings* should be considered as either rising or dropping in the direction of the *air flow* or towards the *government lock*.

When using allowances of 1″ for the *lock* and 7/8″ for the *flange*,

32

Example #2

which totals 1-7/8'', and this amount has to be added to a given *stretchout*, this can be done without having to add the figures on paper. As an example: assume the *stretchout to be 34-9/16''*. Mentally add 2'' to the 34'' and subtract 1/8'' from the fraction, and the results are 36-7/16''.

STRAIGHT TRANSITION
Example #2.

The **transition fitting** referred to as Example #2 is shown in Fig. 2-3 and, although this *fitting* is practically the same as Example #1, there are two distinct differences between the two *fittings*. In this Example #2, the opening change is from a *square duct* to a *rectangular duct* and the amount of *offset* is quite extreme.

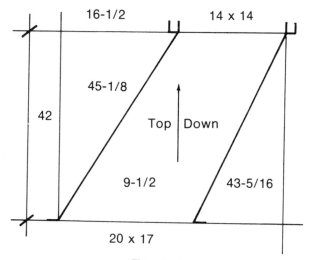

Fig. 2-3

Although this Fitting is not particularly difficult to develop, it is most often laid out incorrectly or marked wrong. Due to its extreme offset, its development does require careful attention. The dimensions used for this particular *example* were taken from an actual *work order* and the *true lengths* have been taken from a Chart.

As shown in Fig. 2-3, the required finished *height* or *length* of the *fitting* is 42'', with a given offset of 16-1/2'': also the *top* drops down 9-1/2''. Aside from the opening sizes at each end, this is all that is required to develop the *fitting*, as the other figures are easily determined. It is the critical figures *only* which are always shown and the *layout* is developed accordingly.

Transitions

The *elevation sketch* shown in Fig. 2-4 should be considered a must for this type of *transition fitting.*

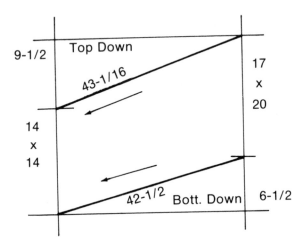

Fig. 2-4

The *fitting* is made in four pieces, described as the *top, bottom, long* and *short sides,* with their *true lengths* being taken from a Chart, and are as follows:

PIECE	CHANGE	TRUE LENGTH
Top	down 9-1/2″	43-1/16″
Bottom	down 6-1/2″	42-1/2″
Long Side	offsets 16-1/2″	45-1/8″
Short Side	offsets 10-1/2″	43-5/16″

With this *fitting,* the *sides* are the best choice for the *pittsburg* edges, with the 1/4″ *single edges* on the *top* and *bottom.* The pieces would be calculated for the *shear list* in the usual manner and would be listed as follows:

Top	31″ x 44-15/16″
Bottom	31″ x 44-3/8″
Long Side	25-1/2″ x 47″
Short Side	25-1/2″ x 45-3/16″

STRAIGHT TRANSITION
Example #3.

The **transition fitting** referred to as Example #3 is shown in Fig. 2-5 as it might appear on a *work sheet.* This type of *fitting* is usually made up in three pieces. However, under certain conditions described

34

Example # 3

here in Example #3. the *fitting* can be made in two pieces.

The basic *rule* that can be applied in determining if the two piece *layout* can be used, is as follows:

RULE: Whenever the dimensions of one opening add together to the same amount as the opening at the other end of the fitting, and either top or bottom and one side is flat, this two piece layout can be applied.

The example shown in Fig. 2-5 is the easiest to recognize, where the dimensions are the same, only reversed. However, this two piece *layout* is not only limited to the same dimensions being at each end. The *rule* only requires that each end adds up to the same amount. Either opening size could be changed, even at both ends, but it must be made so that they each retain the same total.

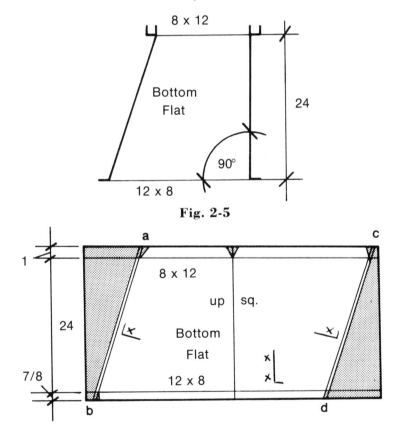

Fig. 2-5

Fig. 2-6

Transitions

As an example: Referring to Fig. 2-5, if the top dimension was kept at 8 x 12, the bottom dimension could be changed to any other size from 19 x 1 on down, such as 18 x 2, 17 x 3, 16 x 4, etc. In each instance, the difference between the two dimensions is equal. This keeps the outside *miter* lines the same, as shown at *a-b* and *c-d* in Fig. 2-6. The openings at both ends can be changed, and the figures do not have to be alike, only add together to the same amount.

To make this better understood, assume that the top opening was 7 x 13 and the bottom opening was 5 x 15. The difference here is 2. Assume again that the top was still 7 x 13 but the bottom was 15 x 5, the difference is 8.

In either case, all that changes is the *layout* pattern shown in Fig. 2-6 which, in turn, would change the *miter lines a-b* and *c-d*. These lines would either shorten or be longer, depending on the change in the opening sizes.

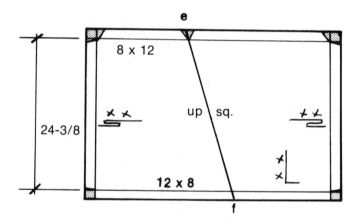

Fig. 2-7

As shown in Fig. 2-7, the width across would be, in this example, 20″ plus 2″ for the *pittsburg locks*. The height would be equal to the length of the *miter* lines *a-b* or *c-d* which are the same, plus the allowance for lock and flange.

To determine the blockout size for the *shear list*, the given finished height of the *fitting* is *triangulated* by the difference between each opening dimension: in this example, showing 8 x 12 to 12 x 8, the difference is 4.

The break line *e-f*, shown in Fig. 2-7, will at **all times** be diagonal on the *layout*, but this is no problem. The flange can be flattened

Example # 4

out enough to allow for the diagonal break and can easily be straightened out again with a hand *tongs*.

This *fitting* can be developed to have either the top or bottom flat, as it is only a matter of placing the markings on the opposite side of each piece of the *layout*.

The same type of *fitting*, as shown here, is often required with a rise or drop on either the top or bottom. This then makes the development of this *fitting* similar to Examples #1 and #2 of this Chapter. The flat side of the *fitting* shown in Fig. 2-5 would have to offset down or up, as required, and the development would have to be handled as a four-piece *fitting layout*.

The blockout sizes are determined in the usual manner. However, the *miter lines* on the flat side of the *fitting* are the *true lengths* of the top and bottom.

STRAIGHT TRANSITION
Example #4

The **transition fitting** referred to as Example #4 is shown in Fig. 2-8 and is perhaps the easiest of all transition *fittings* to both lay out and fabricate. This *fitting* is shown as it might appear on the *work sheet*, although this same *fitting* is often viewed from the straight side.

This *fitting* is practically always made in two pieces. However, when the size of the *fitting* becomes large enough to require the

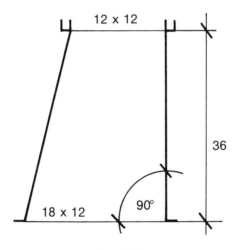

12 x 12

36

18 x 12

90°

Fig. 2-8

Transitions

use of heavy metal, the *fitting* should be made in four pieces.

When size permits, the *fitting* should be made with a three-sided section having *pittsburg* locks on two sides. Fig. 2-9 shows the shaded area representing the cut-away material. The drop-in panel is shown in Fig. 2-10 and its length, 36-1/2″, is the determined true length triangulated from Fig. 2-8.

Here, again, is one of the little points that is often overlooked. Placing the *pittsburg* on the three-sided section makes it easier

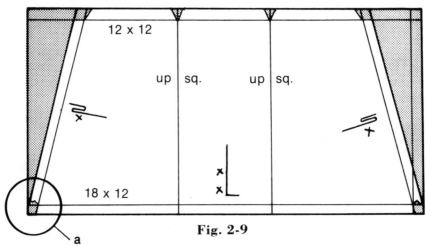

Fig. 2-9

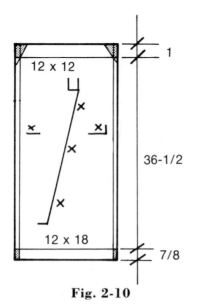

Fig. 2-10

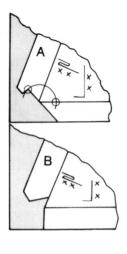

Fig. 2-11

Example #5

to form at the *brake*. It is decidedly easier to fabricate, particularly when the *fitting* is of fairly large proportions.

The encircled notch, *a* in Fig. 2-9, is shown as two corner blowups, *A* and *B* in Fig. 2-11. Notching *Method A* can be figured quite accurately, as shown by the small circles, indicating just where the lock will finish up after it is run through the machine. Exact measurements are not necessary since the *cutter* can make a fairly close guess. *Method A* is preferred by the *author*. However, many *cutters* use *Method B*, finding it easier to sight the 3/8" cut on the line, Fig. 2-11. Either *method* is correct since both will serve the purpose.

The material should be ordered on the *shear list* as follows:

Fig. 2-9   37-7/8" x 50"
Fig. 2-10   38-3/8" x 12-1/2"

STRAIGHT TRANSITION
Example #5.

The **transition fitting** referred to as Example #5 is shown in Fig. 2-12 and is typical of the large size *fittings* which are used for connecting *duct work* to *fan room* equipment, such as **coils, blowers, de-humidifiers,** etc.

This type of *fitting* is usually of such large proportions that it often requires several sheets of metal to make up just one side sec-

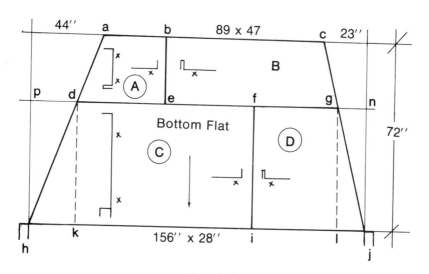

Fig. 2-12

Transitions

tion. Heavy metal is always used for these *fittings*, ranging anywhere from 22 to 16 gauge, according to the *job specifications*. Due to the heavy metal and size of these *fittings*, their construction is quite similar to *fan room housings*.

The same fact applies here as with other *transitions* in this Chapter; that no one specific *layout* could be given that would serve as an all purpose *pattern*. However, the ability to handle each phase in developing the *example* shown in Fig. 2-12 will enable the *cutter* to handle any similar type of large *fitting*, regardless of the variations in its design.

Theoretically, there is **no** difference in the *layout* of the small or large *transition fitting*, as both are developed by *triangulation*, and the same facts and figures have to be determined in either case. However, whenever the *fitting* is of such extreme proportions that it cannot be developed within the width and length of a full sheet of metal, its development does become somewhat more complex and must be handled in a different manner.

There are actually two *methods* which may be applied in the development of this type of *transition fitting*: using only **mathematics** or working from a **scaled detail**. Either of these *methods*, when applied alone, are usually too slow a process to be considered practical for *shop* use. The *author* prefers using a combination of both *methods*, as it cuts down on layout time and makes it possible to re-check measurements throughout the *layout* procedure.

The *scaled drawing* must be carefully developed or it will not be accurate enough to permit transferring measurements from it, directly onto the full size *layout* being developed. The 4 foot steel **circumference bench rule,** which is used in all *shops*, is excellent for developing the *scaled drawings* or *detail*. It is referred to by either *term*. Use the 1/8″ *scale* which extends across the bottom edge of the *rule*, with the large numbers on this same *scale* representing 1″ divisions. Although the small divisions are read 1/8″, it is possible to sight 1/16″ readings in the spaces between the 1/8″ divisions, even though they are not indicated on the *scale*. This does, however, require extreme care in laying out each measurement and it is essential that a fine pointed *scratch awl* be used.

As the *scaled detail* is merely reproducing the given measurements from Fig. 2-12, using the *scale rule*, there is no purpose in showing a *scaled drawing* here also. Although Fig. 2-12 is **not** to any specific *scale*, it will serve the same purpose, by referring to the various points which are lettered for reference.

Example #5

When the *scaled detail* is completed, the various pieces should be calculated for the *shear list*, taking advantage of both the *detail* and *mathematics* in preparing same. As the *cutter* becomes familiar with working in this manner, he will find it easier to prepare the *shear list*.

The block-out sizes are handled in the usual manner, adding the necessary allowances as required, therefore no further discussion is needed. It is only the development of the unknown dimensions, in Fig. 2-12, which must be determined and the *method* of doing so.

It is necessary to determine the measurement between points *d* and *g*, and to do this with *mathematics*, the measurements between points *p* and *d* and *g* and *n* also have to be established. These measurements can be taken off the *scaled detail*. However, it is the purpose here to show how to calculate these unknown figures.

Each of these measurements must be calculated separately, as described here. To obtain the length between points *p* and *d*, it is necessary to determine the **rise per inch** of the *miter line, h-a,* which is found by dividing the given amount of offset by the given length. This is then multiplied by the length of the dotted line *k-d* which is the **net** width of *panels C* and *D* or, as shown in this Example #5, 46-1/8''.

44'' divided by 72'' = .611	
46.125 x .611 = 28.1824 or 28-3/16''	Line *p-d.*
23'' divided by 72'' = .311	
46.125 x .311 = 14.345 or 14-3/8''	Line *g-n.*
28.1823 + 14.3448 = 42.5271 or 42-1/2''	

This sum of 42-1/2'' is the total of lines *p-d* and *g-n*. Subtract this from the overall length or opening size *h-j*, which is 156'', and the difference is 113-1/2'', the determined length of the standing seam *line d-g* in Fig. 2-12.

The horizontal length of *panels A, B, C* and *D* should be proportioned out so that the vertical seams *b-e* and *f-i* are not in line with each other, Fig. 2-12.

The lengths can be left to the *cutter's* own discretion and are often determined by the scrap material that might be available. However, *panel C* should be a full sheet, 120'' x 48'' for this Example #5 and the others figured accordingly.

The two *panels, C* and *D*, should be blocked out to the required size, even though both pieces will require a diagonal cut. The other

Transitions

panels can be blocked out slightly longer than required.

Both *side sections* can be made in one piece and also placed on the *shear list*. Their length will be equal to the diagonal *miter lines* a-h and c-j and, since this bottom is flat, their length can be taken directly off the *scaled drawing*. To calculate their length *mathematically*, refer to the *Math Problem No. 5* in the *Math Section* of this *book* (see **Index**).

The top section is developed in the same manner as the bottom, except for the fact that the top rises 19″ and this must be *triangulated* for the correct or true length. Due to the difference in length between the top and bottom, the points *d* and *g* will **not** be the same as on the bottom, when laid out in the flat, even though a 48″ wide sheet of metal is used here also. The width of *panels A* and *B* will also be wider as they have to compensate for the difference in length between the top and bottom.

To determine the lengths of the *miter lines* for the top section, which has a diagonal slope due to the offset of the side section and the rise of the top, apply the following procedure: Add the **square** of the amount of **rise** or **offset** of the two sections which form the **miter line** to the **square** of the specified **height** or **length of the fitting.** Then extract the **square root** from this **sum.** The result will be the length of the diagonal **miter line.**

Having obtained all the figures required to develop the entire *fitting*, the *shear list* can be completed and the *fitting* developed in the usual manner.

VEE SECTION

Most fittings of this type will develop in the same manner as Example #5, however, in some instances, a *vee* section will be part of the *fitting* and this requires additional calculating. This is shown in *plan view* in Fig. 2-13. No dimensions have been used here, as the only purpose is to describe the method of developing this *vee* section.

A *side elevation* of the *fitting* is shown in Fig. 2-14. Note that both the top and bottom rise as indicated by the shaded areas *r* and *s*.

The development of the *fitting* is basically made off the straight center section, *t* in Fig. 2-14, and sections *u* and *v*, the top and bottom, will each require individual measurements. The *vee* is indicated by the dash line and it must be determined as to just how far into the *fitting* point *m* extends, so that it can be laid out on both the top and

42

Example # 5

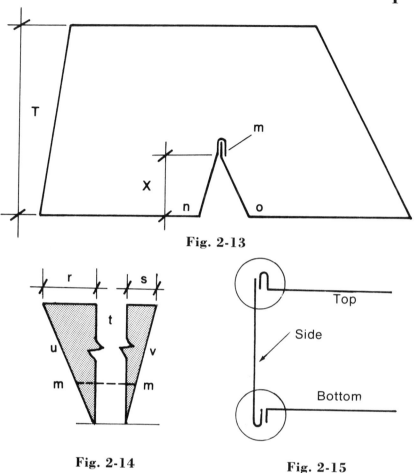

Fig. 2-13

Fig. 2-14   **Fig. 2-15**

bottom sections. Once this has been established, the *layout* of these sections is handled in the same manner as applied in Example #5.

Attention is called to the fact that the *vee* section *miter lines* are developed the same as the *miter lines* on the top section of the *fitting* in Fig. 2-12. Although the *vee* lengths are decidedly shorter, the development is calculated in the same manner. By viewing the *side elevation* in Fig. 2-14, this fact is made quite evident. The outside *miter lines* of this same *fitting* and the *vee* will develop the same way.

In developing the *vee* section, as shown in Fig. 2-13, the points *n* and *o* do not change, which is also true for point *m* in its actual distance from *n* and *o*. This is, of course, in the straight, as indicated by *X*, the same as *T* is the *height* of the *fitting* itself. By measuring off the *height X* in the elevation, these points are established on both the top and bottom sections, Fig. 2-14.

43

Transitions

The vee section should be made in two pieces, with a standing seam as indicated at *m*, Fig. 2-13, in order to use the *pittsburg lock* machine. The standing seam can be made smaller, since its only purpose is to make a seal or joint between the two pieces.

STANDING SEAM CONSTRUCTION.

On some of the very large *fittings*, standing seams are often used to make up the entire *fitting*, instead of the *pittsburg lock*. In these cases, it is usually best to make the double part of the standing seam on the bottom of the *fitting*. This is shown in *elevation* in Fig. 2-15. The usual reason for this construction is the conditions on the *jobsite* where there is insufficient room to install the fabricated *fitting*. This situation is frequently encountered when old buildings are being remodeled. With the *fitting* knocked down, it can be assembled easily by placing the *sides* up and dropping the bottom section in place, from the inside of the *fitting*, if necessary.

STRAIGHT TRANSITION

Example #6

The **transition fitting** referred to as Example #6 is shown in Fig. 2-16 and, although quite similar to Example #5, it is handled in a different manner. Both *fittings* require a standing seam across the top and bottom section. However, in Example #6 the large opening of the *fitting* is within the length of a sheet of metal, so that no vertical seams are required. This more or less serves as a guide in determining when this *layout method* can be applied.

The purpose in showing this *example* is to illustrate how the *miter lines* are developed under suitable conditions. Aside from this, the *fitting* is handled in the same manner as other *transitions* in this Chapter.

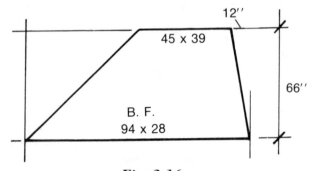

Fig. 2-16

Example # 6

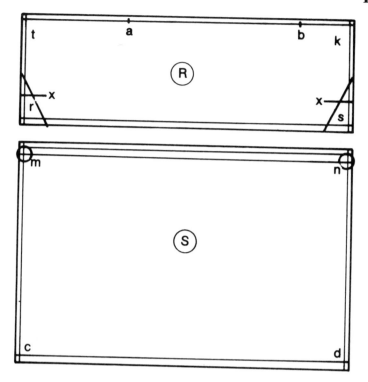

Fig. 2-17

As shown in Fig. 2-17, the two pieces required for the top and bottom will be referred to as panel *R* and *S* with panel *S* being a 48'' wide sheet of material. Adding together 1-7/8'' for the standing seam and 7/8'' for the flange, the total is 2-3/4''. Subtracting this from the width of the sheet, 48'', leaves 45-1/4'' which is then deducted from the finished height of the *fitting* which is 66''. The answer is 20-3/4'' and, allowing 1/16" for the growth of the seam, the **net** length of panel *R* will be 20-11/16''. Add, to this, 1'' for the *lock* and 7/8'' for *flange*, making the blockout size 22-9/16''. The width across the bottom panel *R* should be the same as panel *S*, 94-1/2'', and the *shear list* should read as follows:

2 pc. 94-1/2'' x 48''   *S* bottom & *S* top.
*1 pc. 94-1/2'' x 22-9/16''   *R* bottom.
*1 pc. 94-1/2'' x 23-9/16''   *R* top. (Note that top section is 1''
                           longer).

*Although both the top and bottom panels, *R*, do not require the full 94-1/2'' length as shown, it is decidedly easier and faster to make up the *layout* when the pieces can be lined up together.

Transitions

Referring to Fig. 2-17, panels *R* and *S* should have a 1/4″ edge scribed on the sides. The top of panel *R* should have a 1″ edge scribed across for the *lock* allowance and these lines locate points *t* and *k* where they intersect. Using a 7/8″ scriber, place a line across the bottom of panel *R* to establish line *r-s*.

Scribe a 7/8″ and a 1″ line above it for the standing seam allowance across the top of panel *S* which will intersect the 1/4″ lines already on the panel, locating points *m* and *n*. In the same manner, across the bottom, scribe a 7/8″ edge for flange allowance which will establish points *c* and *d*, Fig. 2-17.

Referring to Fig. 2-17, measure over 12″ from point *k* to locate point *b* and, from *b*, measure over the width of the opening which is 45″ for point *a*. As *c* and *d* are already established in panel *S*, the

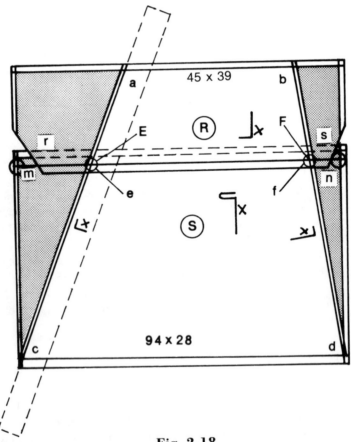

Fig. 2-18

Example # 6

four essential corners of the *layout* are ready. As indicated at points *x* and *x* in panel *R*, make a random diagonal cut-off at each corner. The purpose of this is to permit a view of the line *m-n* of panel *S*, so that the .7/8″ line, *r-s*, of panel *R* can be placed directly **on** line *m-n*, Fig. 2-18.

With the pieces set in the position just described, being certain that right hand edges of both panels are in line with each other, place a *straight-edge* across the two panels so that it is between the four outside corner points *a*, *b*, *c* and *d* of the *layout*.

Being careful not to move the panels, slide the *straightedge* over so that it rests against points *a* and *c* and, with a *scratch awl*, scribe a line to connect these two points. While the *straightedge* is still in this position, using the special 1/4″ scriber made for this purpose and resting it against the side of the *straightedge*, scribe the 1/4″ edge, Fig. 2-18. This will be the single edge for the *pittsburg* seam.

Repeat this same procedure on the other side at points *b* and *d*. Then, prick mark through points *E* and *F*, to also establish these points on panel *S*, which is underneath panel *R*, Fig. 2-18. Both pieces can now be cut out and the usual notching done at points *e* and *f*. The method of notching these corners is shown in the enlarged views in Figs. 2-19 and 2-20.

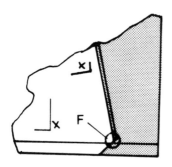

Fig. 2-19

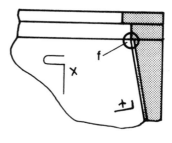

Fig. 2-20

Having determined that the top section is 1″ longer, points *E* and *F* will **not** locate exactly the same as in the bottom section, even though both panels *S* are the same width. Therefore, the same *layout* procedure should be repeated and it should be remembered that while this *layout* was made on the **inside** of the bottom section, it will be on the **outside** for the top section, and it will be necessary

Transitions

to turn the top panels over before marking the fabrication instructions on each piece.

STRAIGHT TRANSITION

Example #7

The **transition fitting** referred to as Example #7 is shown in Fig. 2-21, as it might appear on a *work sheet*. The conventional *method* for developing this same *fitting* has already been described in this Chapter, so no further discussion is required.

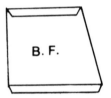

Fig. 2-21

The purpose here, is to show the advantage of laying out smaller *transition fittings*, that require S & D, in one piece and using the *bandsaw* for making multiple cuts from the one *pattern*. As the *drives* are turned after the *fitting* is fabricated, there are no horizontal edges to be turned in the *brake*. Therefore, this *fitting* is formed in the same manner as a section of *duct*.

As shown in Fig. 2-22, the *layout* offers no problems and it saves considerable time and excessive handling of small pieces. The block-out pieces do not have to be cut accurately: however, they should

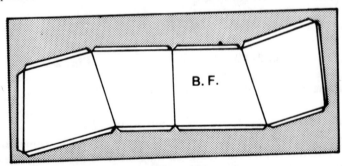

Fig. 2-22

Example # 8

be reasonably close to the determined size, to make it more convenient when stacking and clamping for the *bandsaw*.

The *bandsaw* is the fastest means of cutting multiple pieces, as it only requires the outline being on the top piece which is taken from the original *layout*. The number of pieces that can be cut at one time is variable, depending upon the condition of the *blade* and the speed of the motor on the *saw*. The amount of work that can be cut is usually determined by the thickness of the stack, rather than the number of pieces. This thickness may range from between about 1/2″ to 1″.

The one piece *layout*, as shown here in Fig. 2-22, is not necessarily limited to only very short *fittings*, as this same *method* of *layout* can be applied to any larger size *fitting* that can be developed from a 48″ wide sheet of metal. It should, however, be considered limited in regards to cutting in the *bandsaw*. These larger *fittings* should be cut at the *bench*, cutting one or two pieces at a time with the Unishear.

A stack of larger *fittings* builds up to a weight that makes it too heavy and awkward to be considered practical to attempt to cut in the *bandsaw*. Also, the *throat* of the *bandsaw* limits its capabilities considerably when it comes to cutting any large size pieces. The usual procedure is to cut only about half the *fitting* from one side and then turn it over, change the clamps and trace the *pattern* on the other side. Considering the fact that the *bandsaw* is used to save time, more time would be lost than gained under these conditions.

Unless there is some specific reason for making the large *fitting* in one piece, the *layout* should be made in two or three pieces, using the conventional *method* of *layout*.

STRAIGHT TRANSITION

Example #8

The **transition fitting** referred to as Example #8 is shown in Fig. 2-23, as it might appear on a *work sheet*. This type of *fitting*, which has an extreme amount of offset and overall height, is encountered quite often and, except for one phase of its development, is handled in the same manner as Example #2 of this Chapter.

Since it is just the handling of the *layout* for the top and bottom which is of concern here, the side sections have been purposely omitted. The dimensions shown here point out the extreme difference in the proportions of the *fitting*. As shown in Fig. 2-23, it is not

Transitions

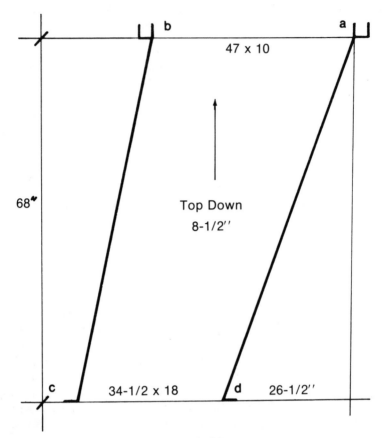

47 x 10

Top Down
8-1/2″

68″

c   34-1/2 x 18   d   26-1/2″

Fig. 2-23

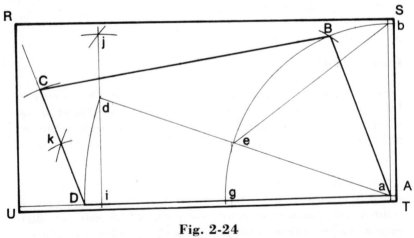

Fig. 2-24

Example # 8

possible to develop the *fitting* in the usual manner, across the sheet, without making a standing seam on both the top and bottom. However, it is possible to lay out the *fitting* lengthwise on the sheet of metal and develop it in one piece as shown in Fig. 2-24.

The method applied here is very simple, being developed by means of basic *geometry*, such as applied in copying any usual given shape. The usual rough *sketch*, shown in Fig. 2-25, should be made, showing the drop of both the top and bottom sections. The straight line height or length of the *fitting* is used in the usual manner to *triangulate* the top and bottom to obtain their **true lengths.**

STEP 1. The rectangle *R, S, T, U*, represents the metal sheet used for the *layout* which is 48 + inches or the width of the sheet as it comes. The length should be a little longer than the *triangulated* length required, Fig. 2-24.

STEP 2. Assume the *fitting* is to have the *pittsburg locks* on the sides, with the 1/4″ edges to be on the top and bottom sections. As the finished top and bottom sections will be off the bottom of the sheet, *U-T*, scribe a 1/4″ line off this edge. At the end, scribe a line

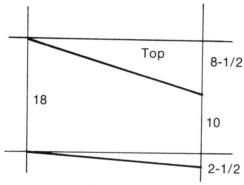

Fig. 2-25

at least 1″ off the edge. The intersecting point of these lines is marked *a*. This will serve as the major point in the development of the layout and, also, the only original point of the *layout* that will be part of the finished *layout*, being identified as point *A*, Fig. 2-24.

STEP 3. Measure up from point *a*, the width of *b-a*, Fig. 2-23, which is 47″ and mark point *b*, Fig. 2-24. Using *trammels* set to the width of *a-b*, scribe a 90° arc to locate point *g*.

STEP 4. From point *a*, measure over, along the base line, the *true length* of the top and establish point *i*; then, squaring up from point *i*, raise a perpendicular line to the top of the sheet, Fig. 2-24. On

Transitions

this same perpendicular line, up from point i, measure off the amount of the offset, 26-1/2" as shown in Fig. 2-23, marking this point d. Using a *straightedge* placed against points d and a, draw in line d-a, Fig. 2-24, which will also establish point e on the arc.

STEP 5. Set *trammels* to the width of *chord e-b* and, with this setting from point g, intersect the arc to locate the new point b which is identified by the capital letter B.

STEP 6. Set *trammels* to the width of span d-a. Using this radius from point a, scribe an arc downward to establish point D. Using e as radius point and with the *trammels* set to a random length that will be near the top of the sheet, scribe an arc to intersect the line up from point d to establish point j, Fig. 2-24. With the *trammels* still set at the same setting and with point g as radius point, scribe an arc which will become point k, following the next step.

STEP 7. Set the *trammels* to the distance between points d and j and, with point D as radius point, scribe an intersecting arc to establish point k. Using a *straightedge* from point D through point k, draw a line up to the top of the sheet, Fig. 2-24.

STEP 8. Setting *trammels* to the distance between points c and d in Fig. 2-23 (opening size of the *fitting*) and with point D as radius point, scribe an arc intersecting the line extending from point D through point k to establish point C, Fig. 2-24. Using a *straightedge* placed against points C and B, draw a connecting line to complete the *layout* of the *top*, except for the necessary allowances for the *lock*, *flange* and 1/4" single edge.

As this *layout* for the *top* has been developed from outside *plan* view, the finished *top* **must** have the markings placed on the under side. The *bottom* will be developed in the same manner, however, the *layout* is being made on the **inside,** therefore, the markings are placed on the same side.

chapter III

TRANSITION ELBOWS
THREE-WAY TRANSITION

There are numerous ways of laying out the **transition elbow;** in fact, few fittings can be handled in so many different ways. Because of this, the *cutter* can select the method that he finds easiest to apply and, at the same time, proves to be most reliable.

The *method* shown here is the one that the *author* uses in his **regular** layout and which can be applied to even the most complicated *fittings.*

Also shown here are two **short** methods for making the same *fitting* that will, if carefully applied, give surprisingly accurate results.

The following step-by-step procedure is for laying out the *elbow* at the bench, directly on the metal, as is done in actual *shop* practice.

REGULAR METHOD

STEP 1. Start the *layout* of the flat cheek by blocking out a piece of metal at the *shear.* While the metal can be roughed out larger than required, make certain one corner is perfectly square.

Scribe a 1" edge along the right-hand side for the lock and a 7/8"

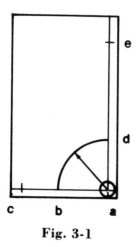

Fig. 3-1

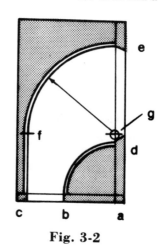

Fig. 3-2

Transition Elbows

edge across the bottom of the sheet for the flange. The intersection of these two lines establishes point *a* in Fig. 3-1.

STEP 2. Using *a* as the radius point, scribe the arc *b-d* to establish the throat. From points *b* and *d*, use the opening sizes of the cheek to locate points *c* and *e*.

STEP 3. Using the length *c-a* from point *e*, intersect the vertical line to locate point *g*, as shown in Fig. 3-2. Keeping the *trammel* points at the same setting and using *g* as radius point, swing an arc down from *e* for the heel.

STEP 4. Using *c* as a radius point and with distance equal to line *a-g*, intersect the arc to establish point *f*. Connect points *f-c* with a straightedge to complete the heel, Fig. 3-2.

NOTE: Figures 3-1 and 3-2 show starting steps for the layout of the cheek, made on the blocked-out metal. This gives the reader a clearer picture of the beginning steps in the layout process. However, the following figures, starting with Fig. 3-3, show the progressive stages of the layout in outline, so that many lines can be omitted from the drawings.

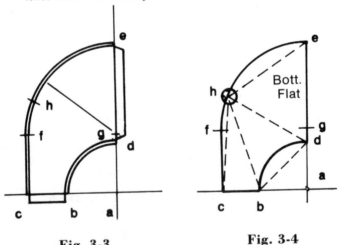

Fig. 3-3   Fig. 3-4

STEP 5. The completed **flat cheek** is outlined in Fig. 3-3. Should the required *fitting* have a flat top or bottom cheek, this *pattern* will be used. If the *fitting* is symmetrical, then the cheek can only be used as a *pattern* for *layout* of the *fitting*, since the cheeks must be the same (except only made as a left and right).

STEP 6. Locate the center of the heel, shown at *h* in Fig. 3-3 using a *flexible rule*. The five broken lines shown in Fig. 3-4, represent the lines to be *triangulated* in order to *layout* the new cheek which is often referred to as the **warped** cheek. In actual *layout*, the broken lines are omitted. They are included here only to help the *reader* visualize the *layout* procedure.

STEP 7. Referring to Fig. 3-5, only two lines are *triangulated* against the full *rise;* one of these being the throat radius, which will be explained later. The other line is shown in Fig. 3-4, *b-d* across the throat. All of the others are *triangulated* against one-half of the *rise.* It is not necessary to lay out the heel and throat at this time, as they play no part in the actual *layout* of the *cheeks.*

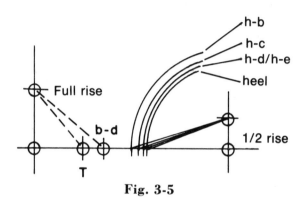

Fig. 3-5

STEP 8. The usual *triangulating* method is now used in developing the **warped** cheek. As shown in Fig. 3-8, points *C* and *B* remain the same as in the flat cheek. However, for a clearer understanding of

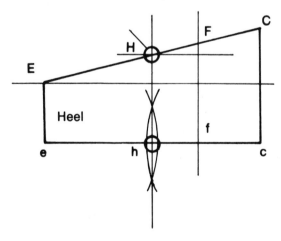

Fig. 3-6

Transition Elbows

this development, the **small** letters shown at each point on the **flat** cheek will be shown as **capitals** on the **warped** cheek, Fig. 3-7.

STEP 9. Also note in Fig. 3-8 that line E-D is the same length as it is on the flat cheek. The development of the lines is started from points C and B, with the triangulated or true length lines taken from Fig. 3-5. These are used in the order of 1, 2, 3, 4, 5, and 6 as shown in Fig. 3-8. No. 6 is the known line E-D and is the actual opening size of the cheek.

STEP 10. With the *square* set at points C and B, locate point F as was done in Figs. 3-2 and 3-9. However, to obtain accuracy in establishing point F, lay out the heel as shown in Fig. 3-6. By squaring up from one line to the other, the exact length of F-C is found.

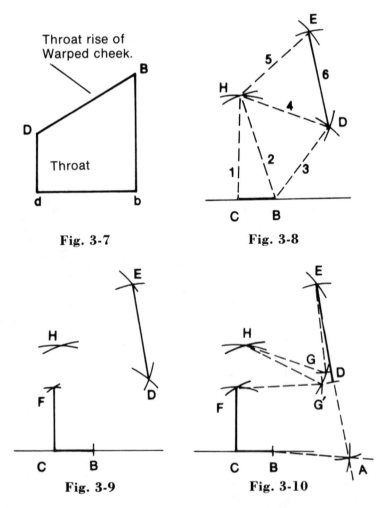

Fig. 3-7

Fig. 3-8

Fig. 3-9

Fig. 3-10

STEP 11. Locate the radius points using the *triangulated* radius lengths developed in Fig. 3-5. From points *H* and *E*, Fig. 3-10, intersect arcs to establish point *G*.

In the same manner, this time from points *H* and *F*, establish point *G'* using the same length of radius in both cases.

NOTE: Due to the straight section of the heel, the short arc between H-F in Fig. 3-11, does not rise 1/2 as does the arc from H-E. Therefore, using the point G' is incorrect. However, since the arc is so short, it is of minor importance. If necessary, shape any bulge freehand. Should the correct radius be desired, it can be easily triangulated using the correct rise determined from Fig. 3-6.

STEP 12. Establish the throat radius point *A*, scribing the arc from *B-D* using the *triangulated* length of the throat radius *a-b*, Fig. 3-5.

STEP 13. Take a measurement of the heel and throat as a precautionary measure. Add all necessary edges, marking as required. The difference between the cheeks is shown in Fig. 3-12.

STEP 14. Mark in the two lines from *H* to *bB* and *H* to *D*, showing which of the lines break up or down and indicate how much.

The break is marked kink up or kink down, as required. To be more explicit, mark for either a light or heavy kink, since a specific degree of break is not necessary.

STEP 15. Complete the **flat cheek** by adding the necessary edges and flanges, cut out, and mark.

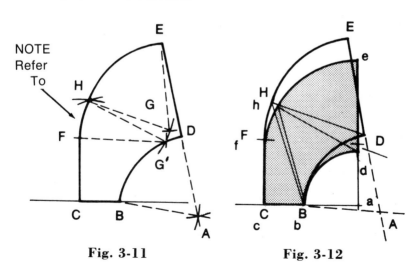

Fig. 3-11   Fig. 3-12

Transition Elbows

STEP 16. The *cutter* usually has the **heel** and **throat** blocked out along with his regular *shear list* and their development is handled in the usual manner. For the heel, determine the *stretchout* of the radius *e-g* shown in Fig. 3-2, plus the span from *f* to *c*, also shown here. For the throat, determine the stretchout of radius *a-d*.

SHORT SLIDE METHOD #1

This *method* can be used in practically all cases. However, when the fitting in question has an extremely sharp rise, the **regular** *method* of *layout* is better.

STEP 1. Lay out the **flat cheek** in the usual manner, as outlined, *a,d,e,g*, in Fig. 3-13.

NOTE: Although this cheek layout in Fig. 3-13 is shown in outline, it must be fully completed before it can be used in developing the warped cheek. Fig. 3-14 shows the flat cheek, complete with the necessary edges and notching, ready to be used as a pattern.

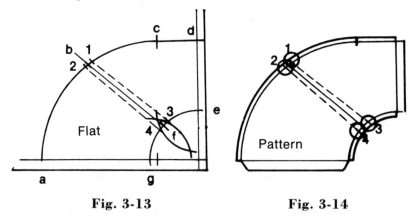

Fig. 3-13   Fig. 3-14

STEP 2. Use *dividers* to find the center of the throat, point *f*, and a *flexible rule* to locate the center of the heel, point *b*, in Fig. 3-13.

STEP 3. Lay out the heel and throat *wrappers* in the usual manner. Measure the length of the newly determined line, *x*, of the heel, which is shown in outline in Fig. 3-15.

Subtract from this length the flat *stretchout* taken from the heel *a-d*, Fig. 3-13, which is also shown as *w* in Fig. 3-15.

STEP 4. In the same manner, find the difference between line *y* of the throat, Fig. 3-16, and the flat stretchout, *g-e*, from Fig. 3-13.

The shaded areas in Figs. 3-15 and 3-16 merely indicate the actual growth, having no purpose or part in the *layout* procedure.

STEP 5. Having determined the difference in both the heel and throat, which is the total amount of rise, refer back to Fig. 3-13. At point *b* of the heel, measure off one-half of the total *rise*, establishing points *1* and *2*.

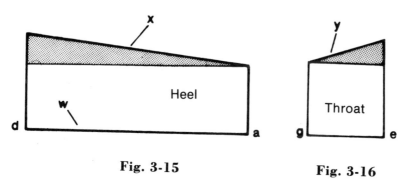

Fig. 3-15   **Fig. 3-16**

STEP 6. Follow through in the same manner for the throat from point *f*, establishing the points *3* and *4*, as shown in Fig. 3-14.

STEP 7. Having placed the required allowances on each side of the center line *b-f*, Fig. 3-13, it is the author's practice to connect these points with lines, as shown. These points are identified in Fig. 3-14 as *1,2,3,* and *4.*

STEP 8. Add the necessary edges for lock, flange, etc. Then cut out, as usual, to complete the flat cheek, as shown in Fig. 3-14. This serves as the *pattern* for developing the **warped cheek.**

STEP 9. Select a piece of metal considerably larger than the flat *pattern* of the cheek. Place a line at random, *j-k* in Fig. 3-17, allowing enough room on the sheet to rotate the *pattern* in both directions.

STEP 10. On the line *j-k*, locate point *m* which is the starting point of the cheek *layout*.

STEP 11. Fig. 3-18 shows the *method* for developing the **warped** cheek. Diagrams A through D are assumed to be lying directly on the sheet of metal described in Step 9 and shown in Fig. 3-17.

STEP 12. Diagram A shows the **flat cheek** pattern in the starting position, with point *b* set directly on point *m* of line *j-k*.

Securely holding point *b* of the *pattern* with a *scratch awl* or some other sharp tool, rotate the *pattern* downward, as indicated by the

Transition Elbows

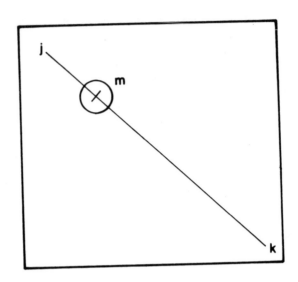

Fig. 3-17

arrow shown in diagram B, so that point *3* sets on line *j-k*.

STEP 13. Now hold point *3* on the line and rotate the other side downward, as indicated by arrow in diagram C, until the point *1* also rests on line *j-k*. Line 1-3 is now directly on line *j-k*.

Scribe the entire contour of the lower half of the *pattern* onto the metal, starting from point *1* and following around through points *2* and *4*, until point *3* is reached. This outlines the lower half of the cheek directly on the metal.

STEP 14. Return the *pattern* to its original position as shown in diagram A. Again holding firmly at point *m*, rotate the pattern upward, as indicated by the arrow, until point *4* sets on line *j-k*, as shown in diagram D.

STEP 15. Holding point *4* firmly on line *j-k*, rotate the other end upward, as indicated by the arrow, until point *2* sets on same line *j-k*. This is shown in diagram E.

STEP 16. Scribe the entire contour of the upper half of the *pattern* onto the metal, starting from point *2* and following around through points *1* and *3* until point *4* is reached. This outlines the upper half of the cheek directly on the metal to complete the **warped** cheek layout.

STEP 17. Note in diagram F, indicated by the arrows, that in both

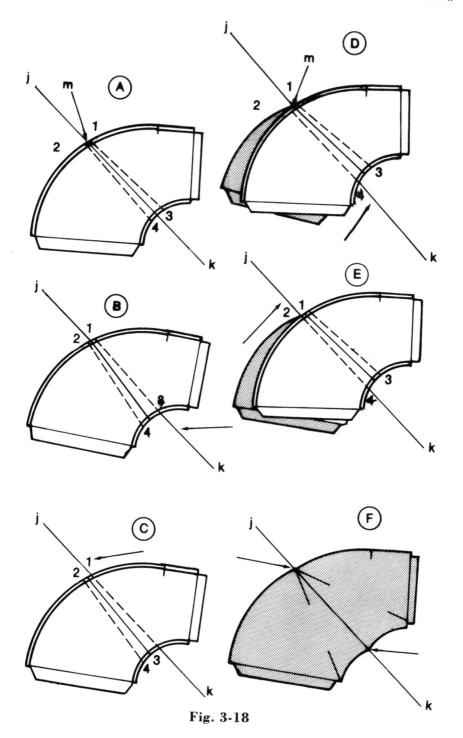

Fig. 3-18

Transition Elbows

the heel and throat the curve or radius is slightly irregular where the arcs intersect.

This is **not** an *error* in the *layout* but a condition that occurs in any *fitting* that is laid out by the *short method*. This is easily rectified by drawing in the short lines by hand to smooth out the curves of both the heel and throat*.

This must be taken care of before attempting to cut out the **warped** cheek.

NOTE: The elbow layout described here refers to a fitting having either a flat bottom or flat top.

When the fitting is symmetrical, only 1/2 of the total *rise* must be taken up on each side. Therefore, only 1/4 of the *rise* of growth is placed on each side of the *center line*.

When the **warped** cheek is completed, it is necessary to make a second one, discarding the original flat cheek that was used in the *layout*.

*It is a good practice to measure out the length of both the heel and throat, to make certain that the cheek will fit. There is always the possibility that too much metal will be cut away while straightening or smoothing out these lines thus making the edges too short. The best practice is to check out any fitting before considering it to be complete.

SHORT METHOD #2

This method of *layout* is not recommended for *elbows* having too drastic a change, or where accuracy is required. However, this is a fast *method* and, if applied carefully, can be used for the *layout* of many *elbows*.

STEP 1. Lay out the **flat cheek** in the usual manner, *a, b, c, d, e,* in Fig. 3-19.

STEP 2. Lay out the heel and throat, determining from them the amount of growth from the flat side to the long side, Fig. 3-20.

STEP 3. Triangulate the heel and throat radius, with the full *rise* being used for both.

STEP 4. Directly above point *c*, and on line *g-c*, raise up the amount of heel growth: thereby establishing point *C*, Fig. 3-21.

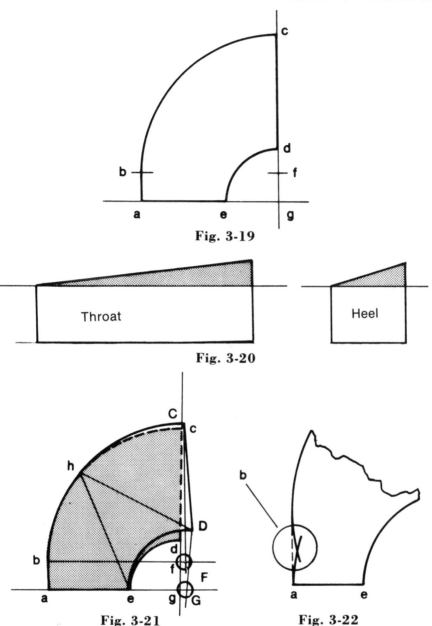

Fig. 3-19

Throat

Heel

Fig. 3-20

Fig. 3-21   Fig. 3-22

STEP 5. From point *d*, scribe an arc equal to the growth of the throat. With the *dividers* set to the opening size, *c-d*, and from point *C*, intersect that arc, thus locating the point *D*, Fig. 3-21.

STEP 6. Using the new throat radius length from points *e* and *D*, intersect arcs to establish point *G*, Fig. 3-21.

Transition Elbows

STEP 7. Using the new heel radius length, and from points b and C, intersect arcs to establish point F, Fig. 3-21.

STEP 8. From these points, respectively, scribe the heel and throat, as indicated by the solid arcs in Fig. 3-21.

STEP 9. Mark the necessary kink up or down; add the edges required to complete the *fitting*.

NOTE: The new heel radius line may possibly show a sharp dip at point b where it joins the short straight section a-b. This can be smoothed by means of a short freehand line, as shown in Fig. 3-22.

SPECIAL DESIGN, UNUSUAL PROPORTIONS

Occasionally, the cutter must develop a *fitting* which has been ordered *special* to fit into a certain spot on the *job*. A good example is the **transition elbow** described here which is taken from an actual *shop work order*. Although the *cutter* may never again encounter this *one-of-a-kind fitting*, he should understand its development, since any similar *fitting* will be handled in the same manner. This *fitting* is a graphic example of what the *author* referred to in the first paragraph on **triangulation** (see **Index**).

As shown in Fig. 3-23, the *elbow* or *angle* is 30°, having a *rise* of 24-1/4" in both the *throat* and *heel*. Due to the extreme *rise* in such a short span the *cutter* must deviate from the usual *method* of *development*. *Triangulation* is still the recommended *method* of *layout*,

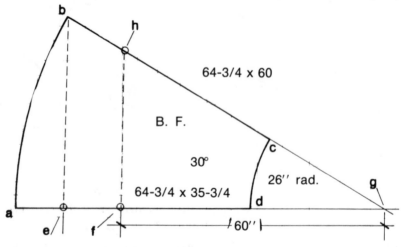

Fig. 3-23

but it is applied in a different manner. This particular type of *fitting* must be handled carefully, particularly when it is to be made out of heavy metal such as the 14 gauge *black iron* used here.

STEP 1. Develop a light gauge metal *pattern* which is the usual practice when working up heavy gauge *fittings*. First, determine if the *layout* of the flat cheek can be developed in one piece. This requires some calculations, using *trigonometry*, to determine if the height *b-e*, which is shown in Fig. 3-23, can be made out of the regular 48" sheet of metal.

Although the same development is explained in the section on *trigonometry*, the *author* feels this is one of the most important uses the *cutter* has for *trigonometry* and it cannot be brought to the attention of the reader too often so that he will become used to applying it whenever possible.

The dash line, *b-e*, shown in Fig. 3-23 represents the **sine function.** Referring to a *trigonometry chart* under 30° *angle*, the *sine function* is .5000 or .5 as it will be used. Add the amount of radius to the width of the cheek and the total length is 90-3/4" or 90.75". Multiply this figure by .5: the result is 45.375" or, as a workable fraction, 45-3/8", which is the height of the line *b-e*, making it possible to obtain out of the 48" sheet.

STEP 2. Select a sheet of metal long enough to be in excess of the total radius and width of the cheek together. Place it on the bench and scribe a line about 1/2" off the bottom clear across the sheet. At a point near the right-hand corner of the sheet and on this line, place a prick mark which will represent the *apex* of the desired *angle* of 30°.

From this point *g*, shown in Fig. 3-23, measure off the 26" for point *d* which is the radius. From point *d*, measure out the width of the cheek, which is 64-3/4", and place a prick mark which will be the radius point of the heel. Again from point *d*, select a random figure (for this particular example use 34") and measure it off from point *d*. This will, when added together with the radius, make a total of 60" and is indicated by point *f*, Fig. 3-23.

STEP 3. In the same manner used to obtain the *sine function*, at the same degree but in the next column which is the **tangent function,** read the figure .5774 and multiply this by 60". This will give the height of line *f-h* as 34.6440" or, as a workable fraction, 34-5/8" which is just 1/64" under this figure. Then measuring up from the base point *f*, being certain to be perfectly square, establish point *h*.

Transition Elbows

Then, using a long *straightedge* resting against points *g* and *h*, scribe a line which should pass directly through point *b*, if all calculations are correct, and this will be the 30° *angle*.

STEP 4. Using *trammel points*, scribe in the *throat* and *heel* to complete the *layout* of the flat cheek in Fig. 3-23.

NOTE: To avoid confusion of too many lines being shown in Fig. 3-23, a copy is shown in Fig. 3-24. This shows the method of developing the various lines which in actual practice would be done in Fig. 3-23.

Since it is best to space out the heel and throat as small as might prove effective in regards to accuracy of the layout, both have been divided into 4 parts. The rise being 24-1/4", this makes it possible to make the *transition* between each opening at 6-1/16" from point to point. The main purpose when considering the spacing is to aid in the forming of the *warped* or *triangulated cheek* which is, in this case, the *top*. The spacing is divided up in the usual manner, locating the center and then dividing this again (Fig. 3-24).

STEP 5. As shown in Fig. 3-29, the *top cheek* is developed, in the usual manner, by *triangulating lines 2-3, 3-4*, etc., against 1/4 of the total *rise* which is 6-1/16". Attention is called to the fact that the dash lines of the *top cheek* are, theoretically, just a shade longer than the solid lines. However, whenever the *fitting* is as large in proportion as this, the difference is not noticeable. Therefore, they are considered to be all the same lengths and the one length is *triangulated* and applied across the *cheek* in the usual manner of development.

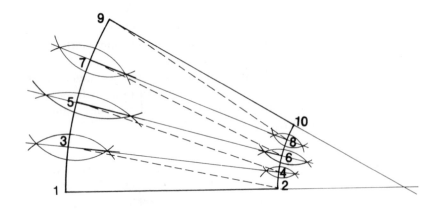

Fig. 3-24

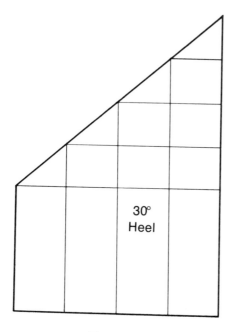
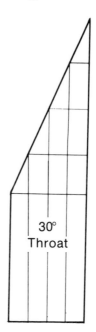

Fig. 3-25   Fig. 3-26

The developed *pattern* will closely resemble the *layout* for a symmetrical *taper fitting* which is quite noticeable in Fig. 3-29.

Referring to Fig. 3-30, length *d* is the amount of growth at each 1/4 point, length *a* is the span across the *cheek*, and the *true length* is indicated at line *e*. Length *b* is the 1/4 division of the *heel* and *c* is the 1/4 division of the *throat*. The diagonal line, which extends from each, represents the *true length* which is applied in developing the *top* or *warped cheek*, Fig. 3-29.

STEP 6. The usual manner of completing the *top warped cheek* is applied and the length of both the *heel* and *throat* should be measured and checked against the top lines of both the *throat* and *heel* in Figs. 3-25 and 3-26.

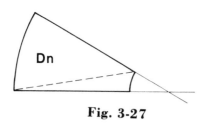
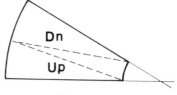

Fig. 3-27   Fig. 3-28

Transition Elbows

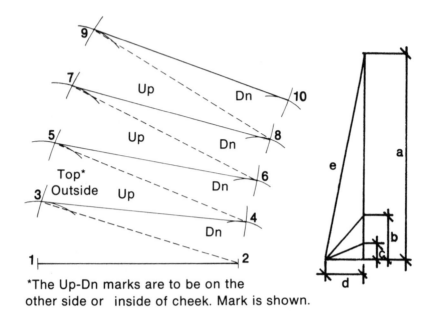

*The Up-Dn marks are to be on the other side or inside of cheek. Mark is shown.

Fig. 3-29   **Fig. 3-30**

STEP 7. The marking of the *top cheek* is quite important and should be given considerable thought. There are three *methods* of shaping the *cheek* shown in Fig. 3-29, which are as follows:

Fig. 3-27, using one break, is not practical, although it is possible to make up the *fitting* in this manner. It creates too sharp a *rise* which will restrict the *air flow.*

Fig. 3-28 is somewhat better, but still does not offer the least amount of resistance to the *air flow.*

Fig. 3-29 shows the best possible *method,* using the points that serve in the development of the fitting, with the up and down breaks being used to raise each of the four divisions of the *throat* up a quarter part of the *rise* at each point, to develop the full amount of the *rise* when at the *heel* point, *9,* and the throat point, *10.*

This is **not** an easy section to form in the *brake* and it is essential that the *brakeman* be able to visualize what must be done. While it is true that both the *heel* and *throat* rise the **same** amount, such as from points *1-9* in the heel and points *2-10* in the throat, it will be found that the *rise* across the *heel* will be flat. When laid on *top* before forming, it can be seen that the *top* will rest on points *1, 9,* and

68

2 of the *throat*, but point *10* will be high up and away from the *throat* section.

Therefore, it is actually a case of breaking the *top* so that it will conform to the *throat*, which makes it necessary to *hit* each division of the *throat* line without losing contact with points *1, 9,* and *2*.

When forming a difficult section such as this, especially when using heavy *iron*, care must be taken to not *over break* any part, as it is difficult to try to bring it back to the desired angle again. A *template* should be made as a guide for this purpose and each kink of a line across the *cheek* should be brought up gradually, as it can be put through several times until the desired shape is developed.

It is the *author's* practice, in regard to a problem of this kind, to form the light gauge *pattern* into the correct shape to fit the *heel* and *throat*. It is always possible that an error was made somewhere along the line. Not knowing this, considerable time can be lost in struggling to make the *top* conform to the right shape, which due to the error might never be possible.

The development of this problem has **not** been to *scale*, but proportioned as considered best for explanatory purposes.

THREE-WAY TRANSITION, WYE BRANCH

The **three-way transition** *fitting* that is described here can be made in any number of different combinations: having either square or rectangular openings, with the top or bottom being flat, or one or both of the side openings 1 and 2, in Fig. 3-31, being symmetrical. Also it is usually proportioned so that the area of opening 3 is equal to or larger than the total size of openings 1 and 2 added together. Although both radii are shown here as being the same, they can also differ from each other. This makes the *fitting* adaptable to almost any conditions encountered.

The *fitting* described here is assumed to have the same radius on both sides, with openings 1 and 2 being the same. The cheek is shown as being flat and can be either the top or bottom when the conditions described here exist. It is only a matter of turning the fitting over to make either the *top* or *bottom* flat.

However, if only **one** opening or radius varies from one to the other and the *fitting* is shown as being *bottom flat* or *top flat*, it is important that the *fitting* be developed accordingly, with the various dimensions being applied as shown on the *work order*.

Three-way Transition

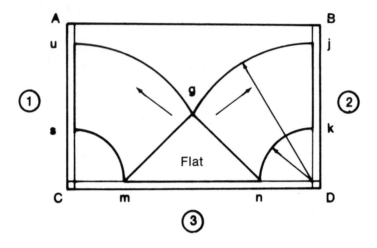

Fig. 3-31

With this type of *fitting*, regardless of the variation that may be between openings 1 and 3 or between 2 and 3, the *triangle* that forms the center *m*, *n*, *g* remains the same height, which is that of the opening *m-n*, Fig. 3-31.

STEP 1. Knowing the size of openings 1, 2, and 3, and the required radius, block out a piece of metal of sufficient size, including allowance for both *lock* and *flange*, as shown in Fig. 3-31. This is indicated by letters *A,B,C,* and *D,* with the width across *C-D* being exact and the height *A-C* to be slightly more than actually required.

Using the given dimensions and radius, lay out the **flat cheek,** establishing points *u,s,j,k,* as shown in Fig. 3-31.

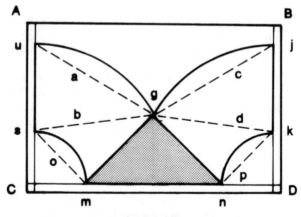

Fig. 3-32

STEP 2. The shaded area of the *triangle*, shown in the drawing in Fig. 3-32 indicates that this area is flat or of the same height throughout.

The broken lines shown in Fig. 3-32 represent the lines that are to be *triangulated.** They are as follows: *a, b, c, d, o,* and *p.*

*For a better understanding of this *method* of *layout*, the *reader* is referred to *Chapter* on *Triangulation* (see **Index**).

STEP 3. The lines are *triangulated* in the usual manner, including the radius lines. Select a piece of metal considerably larger than required and, if the *triangulating* of the lines is to be done on this sheet, E, F, G, H, allow a little more in the width for this purpose. This is the *author's* preference. However, it can be done on a' separate piece of scrap metal.

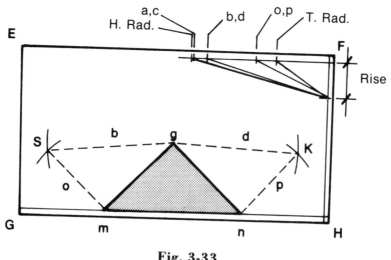

Fig. 3-33

As shown in Fig. 3-33, this calculation is done off the right-hand side of the sheet, although it can be done at either end as preferred. Also note in Fig. 3-33 that the *triangle m, g, n* does not change in any way from the *flat cheek* and is easily transferred to this sheet by laying the *flat cheek* on the top of the piece of metal E, F, G. H and prick marking through the points m, g, n, being sure to keep the bottom edge of the two pieces even. Make certain that sufficient material has been left to develop the *warped cheek*.

Although either side can be developed first, if both sides are the same, as in this example, it is best to work up both sides together.

Three-way Transition

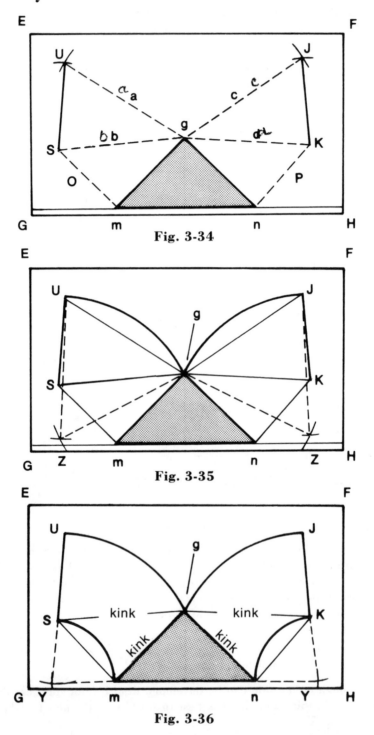

Fig. 3-34

Fig. 3-35

Fig. 3-36

This saves time since the *trammel* settings will both be the same.

Using the true lengths of lines *b* and *o*, scribe intersecting arcs to establish point *S*, as shown in Fig. 3-33. Point *m* is the radius point for the line *o*, point *g* for the line *b*.

STEP 4. In the same manner, using the points *g* and *n*, establish point *K*, Fig. 3-33. Point *n* is radius point for line *p*, point *g* is the point of radius for line *d*.

STEP 5. From point *g*, using the true lines of *a* and *c*, scribe arcs which are intersected with arcs from points *S* and *K*, thereby establishing the points *U* and *J*, Fig. 3-34.

STEP 6. Triangulate the radius lines and, with intersecting arcs from *U-g* and *J-g*, locate the radius points, *Z*, for heels in Fig. 3-35.

STEP 7. Using the same procedure, the throat radius points *Y* are also found, as shown in Fig. 3-36.

STEP 8. Scribing the arcs for the heels and throats and connecting the points *U-S* and *J-K* will complete the *fitting* except for various edges necessary for the *lock, flange*, etc.

STEP 9. The throats can be made in the usual manner. However, it is usually best to make the heel in two parts to avoid making hand *pittsburgs*. Use a small standing seam at point *g*, turned to the inside to tie the heel together.

STEP 10. To properly indicate the break marks on the drop or warped cheek, show the lines *S-g*, *m-g*, *g-n* and *g-K* with a mark of kink up or kink down, as required.

This *fitting* can have a *splitter damper* at point *g* to control the air distribution. However, only use a damper when instructed to do so on the *work sheet*.

The *reader's* attention is called to the fact that the lines *a,b,c,d,o,p*, shown in Fig. 3-32 and those of the radius are *triangulated* in the same manner, using the **total** difference between the openings *1-3* and *2-3*. This applies only when the *fitting* has either the *top* or *bottom flat*. If the *fitting* were symmètrical, which is very often, then only **one half** of the *total rise* or difference would be used in the *triangulating* process. Also, it would require two **warped** cheeks being made, with the *flat cheek* being scrapped.

Although this fitting is used quite frequently, it will be found occasionally that point *g* will drop too low towards opening 3, as shown in Fig. 3-37, making it virtually impossible to fabricate.

Three-way Transition

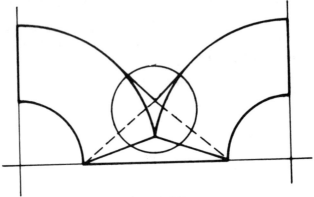

Fig. 3-37

This can be corrected in the *flat pattern* of the *cheek* by carrying two straight lines off the heels which makes the intersection of these lines, or point *g*, easier to make up. The process will still be the same in regards to the development of the *fitting*, both in the *flat cheek* and in the *warped cheek*.

chapter IV

GORED RADIUS ELBOWS

The two *methods of layout* shown here are for very large radius *elbows*, having openings of different sizes, which cannot be cut from one piece of metal.

Following the *methods* shown here, the *pieces* or *gores* can be completely laid out on the bench, **not** only quickly and easily but with dependable accuracy, which is most important to the *cutter*.

For the purpose of keeping each *fitting layout* separate, the two *methods* to be shown in this Chapter will be referred to as follows: **Method No. 1** and **Method No. 2.**

Regardless of the number of *gores* required, the pattern for all the *gores* is laid out on one piece of metal.

For the examples shown, three *gores* are used; namely, *A, B, C,* Fig. 4-1. No scale has been used for this *fitting;* however, it is set to proportions best suited to show the *layout* described here.

SHORT METHOD No. 1

STEP 1. Determine the number of **gores** required. It may be necessary to lay the *fitting* out to *scale* as shown in Fig. 4-1.* This can be

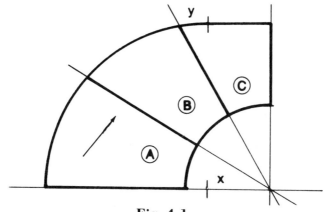

Fig. 4-1

*It is not necessary to scale out the complete elbow as shown in Fig. 4-1; this is for explanatory purposes only.

Gored Radius Elbows

scaled out most conveniently by using 1/8" *scale* on the *bench rule*. This may also be determined with *trigonometry*, which is described elsewhere in this book (see **Index**).

STEP 2. At the base or bottom of the sheet, 13/16" up **(see Method No. 2)**, scribe a horizontal line that is long enough to include the throat and the width of the largest gore, *A*, Fig. 4-2.

STEP 3. Using *e*, Fig. 4-2, as a radius point and with the required radius, scribe in the throat.

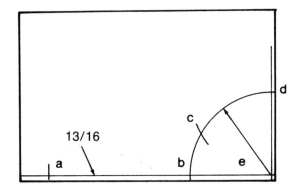

Fig. 4-2

In this *fitting*, with the throat being divided into three parts, the *dividers* are kept at the same throat radius. Then, using point *d* as a radius point, intersect the throat of the *elbow* at *c*.

STEP 4. Draw a straight line through *e* and *c* long enough to reach

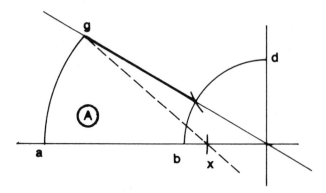

Fig. 4-3

the heel. This will be point g when crossed by the heel radius, Fig. 4-3.

STEP 5. Measure, from point b, the opening size of the largest *gore*, which is A, Fig. 4-1, to establish point a, Fig. 4-3.

STEP 6. Add together the size of the opening on the other end, in *gore* C, and the throat radius.

Measuring off this amount from point a on the baseline, towards the throat, will then establish point x, as shown in Fig. 4-3.

STEP 7. Using point x as a radius point, scribe an arc from point a to g which then completes the *pattern* for gore A, Fig. 4-3.

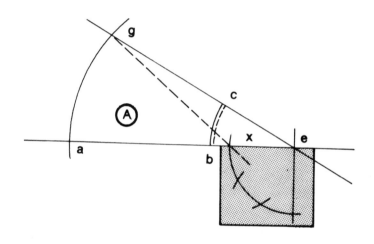

Fig. 4-4

STEP 8. Using e as a point of radius and with *dividers* set to point x, scribe a quarter circle below the baseline. This, of course, will drop below the sheet of metal. Position a piece of scrap metal, large enough to take the required arc, as indicated by the shaded area in Fig. 4-4.

NOTE: The distance between e and x is equal to the straight part of the heel in gore C, Fig. 4-1.

STEP 9. This quarter circle is then divided into the same number of parts as there are gores. In this *fitting*, having three *gores*, divide the circle into three parts. This is easily done by using the *dividers*, still set the same, and dividing the quarter circle shown in Fig. 4-4, thus establishing points j and k, Fig. 4-5.

77

Gored Radius Elbows

STEP 10. Using the **same radius** that was used to draw in the heel of *gore A*, which is the distance from point *x* to *a*, and with *j* as radius point, draw in arc *m* to *l*, Fig. 4-5. This forms the *pattern* for the center *gore B*, shown in Fig. 4-1.

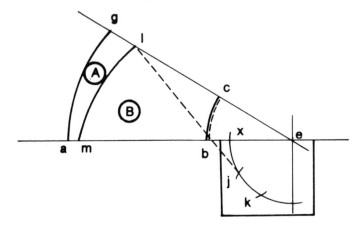

Fig. 4-5

STEP 11. Again using the **same radius** as described in *Step 10*, but using *k* as radius point, draw in the arc from *n*, as shown in Fig. 4-6.

STEP 12. On line *e-g*, from point *c*, measure off the size of the opening of *gore c*, as shown in Fig. 4-1, thus locating point *f*, Fig. 4-6.

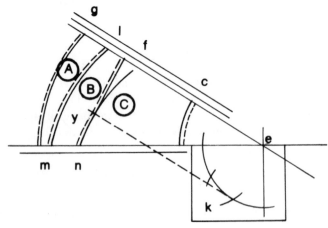

Fig. 4-6

STEP 13. Using *f* as a radius point and with *dividers* set at distance *x-e*, intersect the arc drawn from point *n*, Fig. 4-6.

Form the straight section of *gore C*, as shown in Fig. 4-1, by connecting points *f* and *y* with a straight line, Fig. 4-6.

To check this last step, place a square on line *e-g*, with the center of the *square* set on point *f*. If the *layout* is correct, the three points, *c*, *f*, and *y*, will form a perfect 90° *angle*.

STEP 14. Before removing the scrap metal, draw in the 3/16" edge just back of the *gore C pattern*. This is indicated in Fig. 4-6 by a broken line, part of which is straight.

The 3/16" edge for *gore B* is also drawn in place in the same manner.

STEP 15. Discard the scrap piece of metal. Draw the 3/16" edge in back of *gore A*. This is also shown in Fig. 4-6 by a broken line.

STEP 16. The 3/16" edge is also drawn in the throat, as shown in Fig. 4-6 by a broken line.

STEP 17. Above the diagonal line *e-g*, add 1" + 7/8" for the double part of the standing seam, as shown in Fig. 4-6.

Cut out the whole *pattern* and, before notching, scribe a 7/8" line off the bottom. This line will be the break mark for the flange on *gore A*. It will also be the single edge of the double seam on *gores B* and *C*. By breaking at 7/8", a total of 3/16" is deducted from the overall cheek for growth. Make **two** of the same, right and left, as required for *gore A*, keeping the pattern intact.

STEP 18. Cut away the part of *gore A*, leaving *gore B* and *C* of the *pattern* with the 3/16" on *gore B* intact. Make **two** of these, right and left, still saving the *pattern*.

STEP 19. Cutting away the portion of *gore B*, leave *gore C*, with its 3/16" edge intact for the last *gore*. The 1-7/8" edge which remains on edge *f-c* must be cut down to 1" for lock.

This *pattern* can now be used for the *fitting*, making à second *gore* for a right and left.

STEP 20. To complete the *fitting*, make the throat and heel wrappers in the usual manner.

ALLOWANCE FOR GROWTH IN GORED RADIUS ELBOWS

Both *Method No. 1* and *No. 2* must have some allowance made for growth of material due to the standing seams.

Gored Radius Elbows

The amount of allowance is based on number of *gores*, gauge of metal, and fabrication conditions. The last cannot be accurately allowed for, since too many unforeseen conditions may be present.

The 3/16" allowance made in these *fittings* assumes the use of 22 gauge metal and two seams. With a raw end such as is left for the lock on *gore C*, this allowance is not too critical, since there is a place to trim if necessary. At the same time, if it is too short, the lock will cover the error.

However, the allowance should be figured as closely as possible at all times.

SHORT METHOD No. 2

Although this is presented as *Method No. 2*, it is the method preferred by the *author*.

The actual development of *Method No. 2* is more direct and can be made up completely on one piece of metal, not requiring the added scrap metal needed in *Method No. 1*.

The outstanding difference between the two *layouts* is the fact that, in *Method No. 1*, the horizontal base line is the large opening in *gore A*, with the development of the *fitting* being off this line. In *Method No. 2*, the diagonal line is used for this purpose, with the smaller opening finishing on the base line.

For the purpose of comparison between the *methods*, both *layouts* are taken from the Fig. 4-1. The same letters have been used for

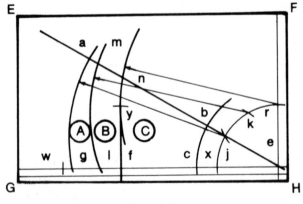

Fig. 4-7

identifying each point, when possible, so that the actual position of the same point can be located easily in each *method of layout.*

STEP 1. In the same manner as used in *Method No. 1,* determine approximate size of material required for the *pattern* and block out at the *shear.*

Let the rectangle *E, F, G, H* represent the blocked-out metal from the *shear,* Fig. 4-7.

STEP 2. Using the scriber for a 1" standing seam, scribe across the bottom of the sheet. At the right side, scribe any desired width, intersecting the seam line to locate point *e,* Fig. 4-7.

STEP 3. The difference between the two openings will determine the radius *e-x* which is used to scribe the arc to establish the points *x* and *r.* With the *dividers* still set, and from points *x* and *r,* intersect this quarter circle to locate the points *j* and *k,* Fig. 4-7. This divides the arc into 3 parts which is equal to the number of *gores* required.

STEP 4. Placing a straightedge at points *e* and *j,* draw a diagonal line across the sheet for the **miter line,** Fig. 4-7.

STEP 5. Set *dividers* to the given throat radius and, using *e* as a radius point, scribe an arc any distance beyond the diagonal line, locating points *c* and *b,* Fig. 4-7.

STEP 6. From point *c,* measure off the width of the largest opening which is in *gore A* and locate point *w.*

STEP 7. From point *c,* measure off the width of the small opening in *gore C* to locate *f.*

With a *square* off the bottom, place a vertical line three-fourths of the way up to the diagonal line. With the *dividers* set equal to points *e-r,* from point *f,* bisect this line to locate point *y,* Fig. 4-7.

STEP 8. Set *trammel* points to the width of *x-w* and, using *j* as radius point, scribe arc *g-a.* Still maintaining the same setting, and using *k* as radius point, scribe arc *l-m.* Then, using *r* as radius point, scribe arc *y-n* as shown in Fig. 4-7.

STEP 9. Growth allowance is made in the same manner as used in *Method No. 1;* however, the allowance is made at the diagonal line. Place a line above and parallel to diagonal line *a-b* at 13/16". This will serve as the cutting line for the flange of *gore A* and the single edge of the standing seam for *gores B* and *C.* To make up the required growth allowance of 3/16", place a line 1/16" below and

Gored Radius Elbows

parallel to the diagonal line *a-b* and prick mark this line for the *brake*, Fig. 4-8.

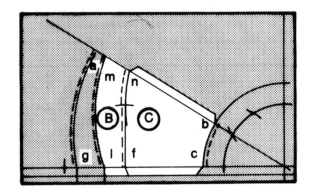

Fig. 4-8

STEP 10. Mark out the notches in the usual manner at points *a*, *b*, *c*, and *g*, Fig. 4-8.

STEP 11. Set *trammels* to 3/16" beyond the actual outlines of gores *A*, *B*, and *C*, and scribe these cutting lines, as indicated by the broken lines. Also, continue the straight line of the heel on *gore C*, Fig. 4-8.

STEP 12. Cut out the entire *pattern*, notching the corners as required, making a right and left for *gore A*. Then cut away the part of *gore A* shown in the shaded area, Fig. 4-8.

STEP 13. Cut out this *pattern* and notch at corners *m* and *l*, as required, making a right and left for *gore B*, Fig. 4-8.

Then cut away this part of *gore B* as shown in the shaded area, Fig. 4-8.

STEP 14. This leaves the final *gore C*, which can be used. Notch out the corners *f* and *m*, in the usual way, and make a right and left cheek.

Mark the cheek *gores*, as required, to complete the cheek *layout*.

STEP 15. The heel and throat wrappers are made in the usual manner. Calculate the *stretchout* for the 90° arc with the radius equal to *x-w*, plus the straight section *y-f*, Fig. 4-7, and an allowance for lock and flange. The throat is, of course, the stretchout of the radius *e-c* plus the lock and flange.

chapter V

SQUARE TO ROUND

INTRODUCTION

Although the **square to round** is not difficult to *lay out*, it is considered to be one of the most intricate *fittings* encountered in the *ductwork* which is used in the installation of *commercial air conditioning systems*. It is also one of the most troublesome *fittings* for many *cutters* who are otherwise very capable *layout* men. This can be attributed to the fact that the *square to round* is one of the *fittings* which is not made often enough by the *cutter* to keep in practice. It is this *fitting*, more than any other, that requires considerable experience in order to become proficient in its development.

The major problem encountered in developing the *square to round fitting* is being able to estimate the approximate size of material it will require to *lay out* the *pattern*. Another problem is selecting the best means of placing the *seams*. This ability can only be gained through actual *shop* experience at the *bench*. Since no two *fittings* will be proportioned exactly alike, no set *rule* can be established which will be suitable for all *fittings*.

It is of utmost importance with the *square to round*, perhaps more so than with any other type of *fitting*, that the *cutter* be able to mentally visualize just how the development will appear on the *metal sheet* upon which he will make his *layout*. This is what enables him to determine how high up from the bottom of the *sheet* the starting points of the *layout* should be placed.

The *square to round* is quite similar to the *round taper fitting* inasmuch as both are quite versatile, being capable not only of offsetting within their own length, but also of making the transition in size between its openings.

The *square to round* is actually a *custom-made fitting*, having been ordered for one specific place on the *job*. Although this, more or less, can be claimed for every *fitting* and piece of *duct* on the *job*, it is the *square to round* which is very seldom, if ever, ordered out before it is actually ready to be installed. This *fitting* is usually the connection between the *blower* and a *line of duct* which has been carried up to the point where it is possible to make the final in-between measurements. On-the-job measurements for both the length

Square to Round

of the *fitting* and the relative position between the *blower* opening and the approaching *duct line* determine whether the *square to round* is to be symmetrical or offset. Although a *flex* connection is usually made at the *blower*, it is vitally important that this *fitting* be made exactly *right*. Since this is one of the very few *fittings* which cannot be altered to fit on the *job* and since it is almost always an exposed *fitting* located in the *equipment room*, it is a case of either the *fitting* being made right or being remade.

The *layout methods* shown here in developing these **square to rounds** are both accurate and reliable. By applying them as described, it is possible to *lay out* any type of *square to round*, regardless of its height, amount of offset, or the size of its openings.

NO. 1   CENTER TAPER OR SYMMETRICAL, METHOD A

Fig. 5-1 shows the **center taper,** or symmetrical **square to round,** in an *elevation* view of the *fitting* as it would appear on the *work sheet.* This is one of the easiest of all *square to rounds* to *lay out,* since it only.requires development of a quarter section.

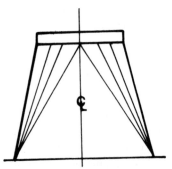

Fig. 5-1

STEP 1. With A, B, C, D of Fig. 5-2 representing the metal for the *layout,* scribe a 1/2" line across the bottom and on the right-hand side of the sheet, with the intersecting point being marked *e.* As the *layout* is to be developed from this corner, scribe another line 7/8" above the bottom line and another vertical line about 2" in from the right-hand side. Mark the intersection of these two lines point *a.*

STEP 2. From point *a,* develop a *plan* view of 1/4 of the *fitting,* with point *c* being the center. With the *trammels* set at the *radius* of the circle and from points *1* and *4,* intersect the circle and establish points *2* and *3,* Fig. 5-2.

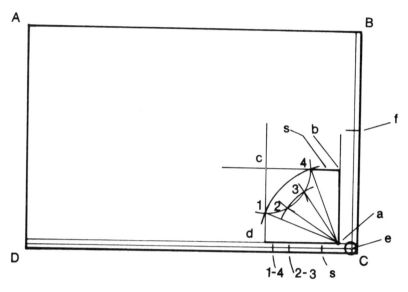

Fig. 5-2

STEP 3. Setting the *trammels* to fit the spans *1-a* and *2-a*, transfer each length onto the 1/2″ line at the bottom, starting from point *e*. With a symmetrical *fitting* such as this one, lengths *1-a* and *4-a* are the same, as are lengths *2-a* and *3-a*. The lengths of line *4-b*, which is the *seam* line, and *1-d* are the same, and these are also transferred onto the bottom 1/2″ line from point *e*, Fig. 5-2.

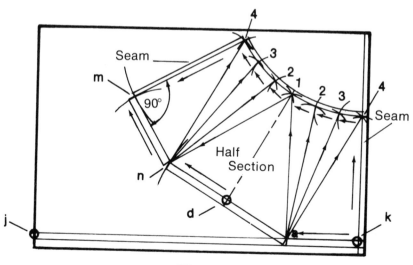

Fig. 5-3

Square to Round

STEP 4. Knowing the required finished height of the *square to round*, not including the *collar*, measure this height up from point *e*, thereby locating point *f*. Using the *trammels* from point *f* to each of the established points on the *line* at the bottom will establish the *true length* lines which are used in developing the 1/2 *pattern* shown in Fig. 5-3.

STEP 5. With a small pair of *dividers* set to one division of the 1/4 circle, the development of the *pattern* is easily completed. The span *a-b* is a *true length*, being equal to one-half of a side at the square opening. This is placed on line *j-k* in Fig. 5-3. Note that point *k* is on the 1/2″ line, directly above point *e*, and that what is now point *k* was point *b* in Fig. 5-2. This means that the 1/4 *layout* shown in Fig. 5-2 is actually rotated a quarter turn in Fig. 5-3 before starting the development of the *pattern*.

STEP 6. With line *a-k* in position, set *trammels* to the *true length* *f-4* in Fig. 5-2. From point *a*, intersect the 1/2″ line at the side in Fig. 5-3.

Merely as point of demonstration, set *trammels* to the *true length* of *f-s*. When placed upward from point *k*, this length should also intersect exactly at point *4*. Likewise, this is the same length as the dashed line, *d-1* in Fig. 5-3, and this is the opposite *seam* line, *m-4*. However, neither of these lines is part of the *layout* at this time.

From this point in the *layout*, the procedure is illustrated in Fig. 5-3, with the various true length lines placed in position. The arrows are used to show the direction in which each line goes and this, in turn, illustrates, better than any possible description, its point of origin.

NOTE: It is the author's conjecture that, since it is one of the easiest square to rounds to lay out, the experienced cutter will not be interested in this particular fitting. Therefore, a special effort has been made to show and describe the development with considerably more elaboration than necessary or required when developing the fitting under actual shop conditions. Although this square to round is simple in design, it contains the basic fundamentals which apply to all square to round fittings of any proportions or design, even though the method of applying them must be handled in a different manner.

STEP 7. The added edge allowance for the *flange* is 7/8″ and for the *acme seam*, 1/2″. Before cutting out the *pattern*, draw in the portion of the 1/2 circle, taking care to have a smoothly flowing line from point to point. Check out the *stretchout* and, if slightly large, leave a little extra allowance when marking off the edge for the *elbow machine*.

FITTING THE COLLAR

With the *square to round*, *taper*, and similar type *fittings* which require a *collar*, the usual width is 2″, with a 1/2″ hem and about 1/4″ allowance for the *elbow machine*, with a single edge on the *collar* and on the *fitting* turn the double.

Use about a gauge heavier material for the *collar*. However, on practically all *fittings* of about 12″ in diameter, it is practical to use 20 gauge. Put two 5/32″ rivet holes for the seam on the *collar*, placing the rivet holes exactly on lines which are the stretchout required, having about 3/8″ to 1/2″ over on each side for lap.

Hem the edge and roll the *collar*. Rivet just one top hole, leaving the other hole for the last operation. Use a clamp or metal screw to hold the *collar* in shape while turning the edge on the *elbow machine*. Also put the double portion of the seam on the *fitting*.

It is rather difficult to hold the spape of the *fitting* when the offset is very extreme or the *fitting* is of large proportions. It is often necessary to make a round disc and place it in the *collar* to retain the shape. Then when fitted onto the *fitting* it will conform to the *collar*. The edge should be peened over in several spots. Before removing the disc, tack-solder in about 6 or 8 places on the inside of the *fitting*. While the disc is still in position, close the edge all around. This should make a tight joint that will hold its shape after the disc is removed.

NO. 2   CENTER TAPER OR SYMMETRICAL, METHOD B

This **square to round** is the same as *Method A* in every way, except that each quarter circle is divided into 4 spaces and the *half-patterns* are developed on a separate sheet. This is due to the fact that this *method* is used more often for the very large *fittings* which require considerably more space. Being larger, they require closer spacing in development of the round portion of the *fitting*.

Select a sheet of metal large enough to just handle the 1/4 *layout* as shown in Fig. 5-4 and indicated by *R, S, T, U*. Proceed as follows:

STEP 1. Scribe a line across the bottom and right side of the sheet, about an inch off the edges of the sheet. Prick mark the intersecting

Square to Round

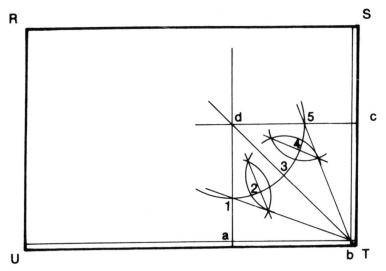

Fig. 5-4

point of these lines and identify as point *b*. Using the given dimensions, establish the points *d, c, and a*.

STEP 2. Set *trammels* equal to the radius of the 1/4 circle and, with point *d* as center, draw in the 1/4 circle to locate points *1* and *5*. This simultaneously establishes lines *c-5* and *a-1*. Either one can

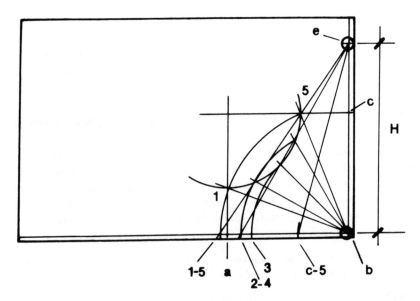

Fig. 5-5

be made the *seam line*. However, in this case, assume that line *c-5* in Fig. 5-5 is to be used for this purpose.

STEP 3. Since this is a symmetrical *fitting*, all four sides of the 1/4 section are the same and a line carried diagonally through points *b* and *d* will locate the center of the 1/4 circle and establish point *3*. Bisect the remainder of the 1/4 circle, using the *trammels* from points *1*, *3*, and *5*, as shown in Fig. 5-4, to locate points *2* and *4*.

STEP 4. Measuring up on the side from point *b* locate point *e* which is the finished height of the *fitting*, not including the *collar*. This height is indicated by *H* in Fig. 5-5.

Using a *trammel*, set the length of line *b-5*. Using point *b* as a center, swing an arc which will pass through point *1*, as it is the same length, and intersect the line along the bottom to establish points *1* and *5*, Fig. 5-5.

STEP 5. Using the same procedure, locate points *2*, *3*, and *4* on the bottom line. Also measure off the seam line, *5-c*, from point *b*. Follow through in the usual manner in developing the half *patterns* shown in Figs. 5-6 and 5-7. Aside from being larger and the *true length* lines being developed on another sheet, no problem should be encountered with this type of development.

STEP 6. Add the 1/2″ on the sides for the *acme* seam and the 7/8″ for the flange. Also check the circumference stretchout and allow

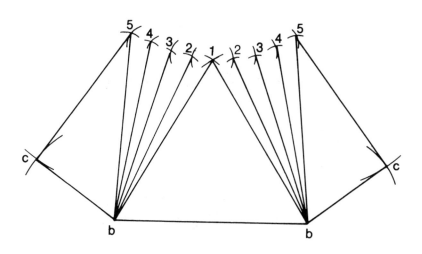

Fig. 5-6

Square to Round

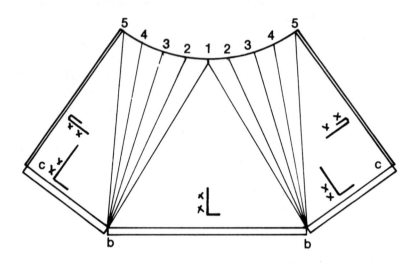

Fig. 5-7

the material for *fitting* the *collar*. The *collar* can be attached by using an *elbow* edge, keeping the double portion of the joint on the *square to round*.

OFF-CENTER

The symmetrical and the off-center **square to round** are both *triangulated* in the same manner. However, considerably more work is involved with the latter since the true length of many more

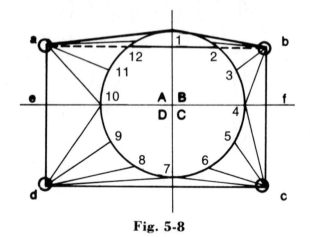

Fig. 5-8

lines must be determined. It will also require a separate or second *pattern* in most cases, unless it offsets only one way and the seams can be set to divide the offset the symmetrical way.

For this or any other off-center *square to round*, it is practical to consider the *fitting* as being in quarter sections, shown as *A, B, C,*

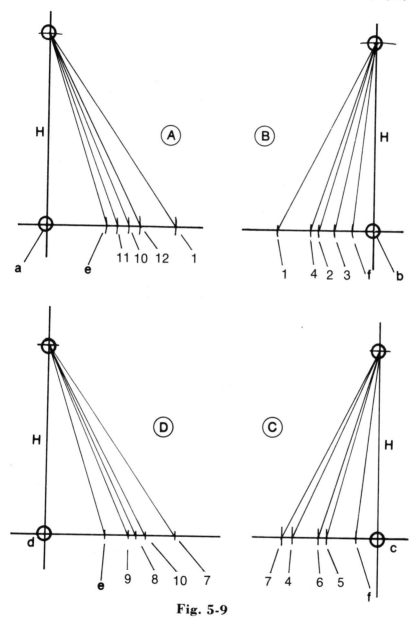

Fig. 5-9

Square to Round

and *D* in Fig. 5-8. For this problem, each quarter must be developed.

Even though only part of the *fitting* may be symmetrical, it is considerably easier to follow through, especially when the *fitting* is laid out and has to be marked for the *brakeman*. It is also easier when the triangulated lines are grouped together, as in Fig. 5-9.

Some *cutters* prefer to develop the *patterns* shown in Figs. 5-10 and 5-11, using each line as it is triangulated and put in its place.

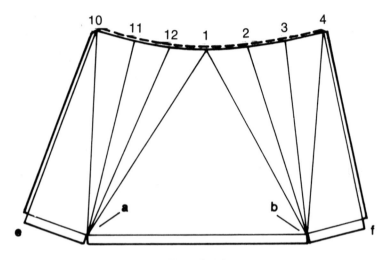

Fig. 5-10

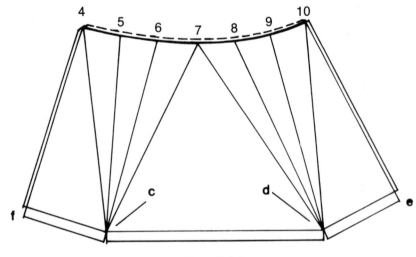

Fig. 5-11

Although this may prove a trifle faster, it may prove to be quite troublesome. If a line is missed somewhere along the way, making it necessary to backtrack, the *cutter* will have plenty of difficulty.

The *author* finds this *method* very good for the simple symmetrical *fittings*. However, when laying out the larger, more complicated, off-center *fittings*, it is best to use the *method* shown here.

Notice that the seam lines, *f-4* and *e-10* in Fig. 5-8 are actually triangulated twice in the groups *A, B, C, and D.* This is, of course, only to illustrate the full set of lines required in developing Figs. 5-10 and 5-11 patterns. In actual shop practice, these are taken from groups *A* and *B*, Fig. 5-9, being triangulated just once.

Having developed the *patterns* shown in Figs. 5-10 and 5-11, the necessary seam allowance and base must be added.

The irregular lines, which will form the circle when the *fitting* is formed, must be carfully drawn in. Even though a *collar* is almost always put on, a neat job is hard to obtain if the raw edge of the *pattern* is too rough or possibly too large.

It is best to lay out the *patterns* as shown. However, when cutting these *patterns*, allow about 1/8'' under the actual development obtained for the *fitting's* round part, shown in Figs. 5-10 and 5-11 by the dotted line. Then, by checking after the *fitting* is fabricated, it is very easy to trim, if necessary, thereby obtaining a clean-cut, accurate edge on which to fit the collar.

On particularly large *fittings* of this type, the spacing along the circular edge is large. For breaking purposes, the spaces should be broken down to about half-inch marks. This is especially true on heavy gauge iron which is used for almost all *large square to round fittings.*

When the seams are to be riveted together, it is best to lay out the rivet holes directly on the seam lines, allowing about a half-inch more stock outside all four seam lines.

RECTANGULAR TO ROUND, ANGLE TO BASE

Because the *round section* of this *fitting* is at an **angle to the base,** the development is slightly more complicated than the regular *square to round.* In order to develop this *layout*, both a *plan* and an *elevation* of the *fitting* are necessary.

STEP 1. Select a sheet of light *iron*, indicated in Fig. 5-12 by the

Square to Round

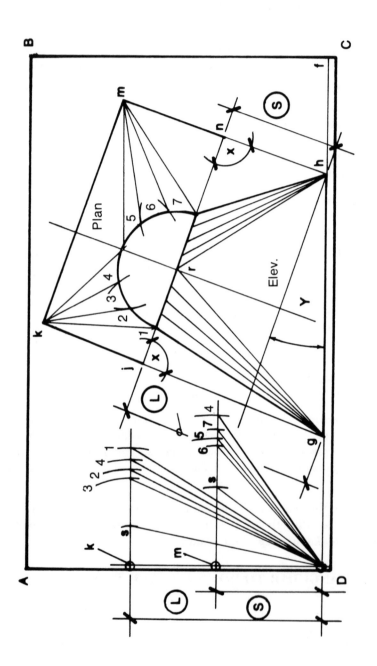

Fig. 5-12

outline *A, B, C, D,* and proceed to *lay out* both the *elevation* and *plan* of the *fitting,* as shown. When the size of the *fitting* makes it impossible to lay out both the *elevation* and *plan* on the same sheet, then it is necessary to *plan layout* on a separate *sheet* of metal. The *elevation* of the *fitting* is only required for the purpose of determining the heights of lines *g-j* and *h-n* and establishing the location of the *round* opening of the *fitting.*

The length of the *base g-h* will also be noted, as this will be the starting point of the *layout.* The degree of *angle Y* must be given with the other dimensions, since this determines the placement of line *j-n* which sets the 90° to each corner of the *base* and is marked as *x* at both corners.

STEP 2. There are several ways of placing the lines which are to be *triangulated.* However, the *author* prefers the *method* shown in Fig. 5-12, where the lengths *L* and *S* are set off, away from the *fitting,* at the side *A-D* of the *sheet.* The outline of the *plan,* shown as *k, m, j, n,* represents one-half of the *fitting.* Scribe the half-circle of the *round opening* from *r* and divide this in the usual manner, establishing points *1, 2, 3, 4, 5, 6, and 7.*

STEP 3. Triangulate the lines *k-1, k-2, k-3, k-4* using height *L.* Also *triangulate* lines *m-4, m-5, m-6, m-7* using the height *S.* These are each placed on their respective lines measuring out from both *L* and *S,* as shown in Fig. 5-12, with their identifying numbers.

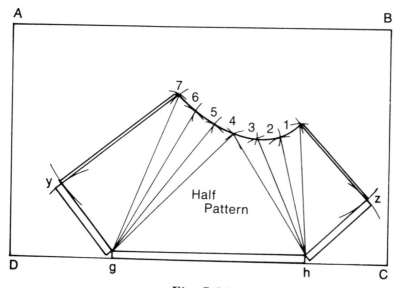

Fig. 5-13

Square to Round

STEP 4. Using a sheet of metal of sufficient size, *A, B, C, D* as shown in Fig. 5-13, scribe a 7/8″ line across the bottom for a flange, which is usually used on this type of *fitting.*

Relocate the line *g-h* onto this line, as shown, and from these points, *g* and *h,* proceed to develop the half-pattern in the usual manner applied to other *square to round fittings.* Lines *y-g* and *h-z* are the width of lines *k-j* and *m-n* which are shown in *plan view,* Fig. 5-12.

STEP 5. The completion of this *fitting* is handled in the usual manner, with added seam allowance and edge for *collar* and *flange.*

Attention is called to the fact that seams *j-1* and *n-7,* shown in the *plan,* are true length lines as they appear in the *elevation,* lines *g-1* and *h-7.* However, they have also been included with the other lines being *triangulated,* points marked *s,* for the purpose of comparison.

LARGE DIMENSIONS USING HEAVY METAL

Figs. 5-14 and 5-15 show the *elevation* and *plan views* of an exceptionally large and very shallow **square to round** fitting. It is usually referred to as a **roof jack** or **flashing** since its purpose is to close the opening in the *roof* between the *curb* and the *riser stack* which comes up through the *roof.*

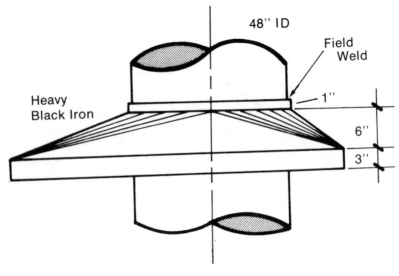

Fig. 5-14

Large Dimensions

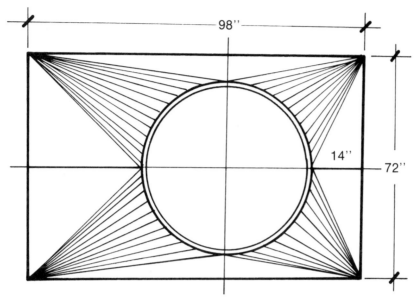

Fig. 5-15

The *layout* of this *fitting* is similar to other *square to rounds*. The purpose for including this particular design is to show how easily it is developed, even though the dimensions are of extremely large proportions.

As shown in Fig. 5-14, the height of the *fitting* is very shallow, 6″ high, with an apron of 3″ to fit over the curb. The opening which fits around the *riser stack* can be fit rather loosely, as it is usually just tack-welded to the *stack* and a *storm collar* is made to fit above this.

The *fitting* is usually made of heavy-gauge black iron. To all appearances, it has every indication of being a very difficult *fitting* to develop. It is, however, quite the opposite and its development will be described and the *layout* shown in Fig. 5-16. Although this particular *fitting* is shown here as symmetrical, this is not always the situation, since the *square to round* is developed to meet the existing conditions. It is not always possible to locate the *riser stack* directly in a center position, as would be preferred. This will not, however, be a major problem, but it will require the development of two *patterns* instead of one, to be used as a *right* and *left*. In the usual manner, a light gauge *pattern* is developed. A sheet of sufficient length and full width should be used, *A,B,C,D* in Fig. 5-16.

STEP 1. In the usual manner, scribe a line across the bottom and the right side, about an inch in from the edges. The intersection of

Square to Round

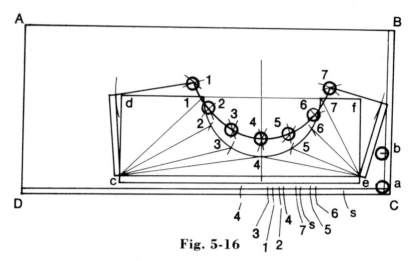

Fig. 5-16

these lines, marked *a*, will serve as the corner of the *triangle* in developing the true length lines.

STEP 2. Establish a horizontal line about 6″ above the bottom of the sheet and use this line to start development of the *fitting*. Allowing room along the right-hand edge for sufficient working area, locate points *c* and *e* from the given measurements of the *fitting*. Also locate the center of the *round* opening in relation to the *rectangular* opening and draw the half-circle.

STEP 3. Divide the half-circle into 6 divisions and number them *1* to 7, inclusive. Now, complete the *layout* by locating points *d* and *f*, Fig. 5-16. In the usual manner, transfer the lengths of the lines, from each corner to the points on the circle, over to the corner at point *a*. From point *a* also locate point *b*, the given height.

STEP 4. Transfer the *seam lines*, *d-1* and *7-f* to the corner at *a*. With the true length developed from point *b* to each numbered point, transfer these lines to the same position. As shown in Fig. 5-16, the *pattern* is superimposed on the original working *layout*. Adding the apron around the sides and developing the irregular curve will complete the *layout*.

NOTE: The dimensions shown here are in no way associated with the layout shown in Fig. 5-16. They are merely random figures used for the purpose of pointing out the fact that this layout is adapted to square to rounds having exceptionally large proportions.

The ringed points shown in Fig. 5-16 represent the new triangulated points of the numbers indicated.

SHORT METHOD, SYMMETRICAL FITTINGS

The *layout method* shown here is good for the short and small dimensioned *fittings* and can be applied to both the *square to round* with an equal square base or one having a rectangular base. The procedure for this *layout* is as follows:

STEP 1. Establish the line *a-b* equal in length to one side of the base or, if rectangular, equal to the longest side. Determine the center point *f* and erect a perpendicular line.

STEP 2. Determine the true height of this side, in the usual manner, and measure up from point *f* the determined length to locate point *e*. At point *e* and parallel to line *a-b*, draw in a line as shown.

STEP 3. Calculate the circumference of the round opening and divide this figure by 4 to obtain a 1/4 length of the circumference. This will be the length of line *c-d* which is centered from point *e*.

STEP 4. Determine the difference between the length of line *a-b* and line *c-d*. Take one-half of this length and, with a pair of *dividers* set to this length, use points *c* and *d* as centers and scribe the two small arcs shown in Fig. 5-17.

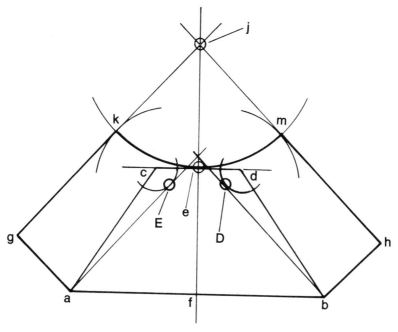

Fig. 5-17

Square to Round

STEP 5. Place a *square* against the arcs at points *E* and *D* and, at the same time, against points *a* and *b*. With the square in this position, scribe the lines *b-h* and *a-g*. Make these two lines equal in length to one-half the width of the *sides* of the base. Each line must be a perfect 90° out from corners *a* and *b*.

STEP 6. Set *dividers* equal to the widths *g-a* and *b-h*, which should be the same, and, with points *E* and *D* as radius points, scribe an *arc* on both sides, as shown. Using a *straightedge* from points *g* and *h*, and resting against the arcs just established, scribe lines as shown in Fig. 5-17. These lines should intersect and bisect the center line at a common point as indicated at point *j*. If accurately developed, the corners *g*, *a*, *b*, and *h* will each be 90°.

STEP 7. With *trammels* set to the length of line *j-e*, and with point *j* as radius point, scribe the half-circle *k-m* which should be measured off to make certain it is the correct length, equal to 1/2 the *circumference*.

STEP 8. Add the necessary allowances for *flange* — 7/8″, *acme seam* — 1/2″, and for the *collar* — about 1/4″ or 5/16″, according to the requirements of the *elbow machine* being used. The completed *half-pattern* is shown in Fig. 5-18, ready for fabrication, which is handled in the usual manner.

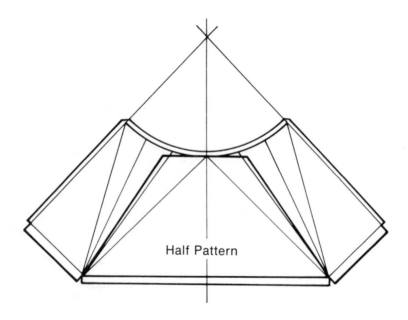

Half Pattern

Fig. 5-18

100

chapter VI

OFFSETS

INTRODUCTION

The following Chapter shows various types of **offsets** developed with both the regular and **short** *methods* of *layout*.

Since the **offset** is one of the most commonly used *fittings*, it is important that the *cutter* be able to apply the correct *methods* of *layout* to the different types he will be required to make at one time or another.

The *layout methods* shown in this Chapter are reliable and can be readily applied in actual *shop* practice.

NOTE: The offsets shown here are numbered for reference purposes only. Where a second method of layout is shown for the same type of fitting, it is described as Method A or B. As previously explained, the author has shown more than one method of layout for the same fitting only in cases where both methods are used to good advantage. It might prove more favorable for the cutter to choose the one he feels he can best apply in actual shop practice.

OFFSET No. 1
Regular with Random Radius, Method A

Although this type of **offset** is not a true radius or *ogee* offset, the capacity is in no way choked. It might even offer less resistance to the air flow than the regularly made, true *offset*. However, it should be used only where the *offset* is clear of any close obstruction, because the straight section may not clear due to the *offset* not rising quickly enough. For this reason, the *cutter* should not attempt to use this *method* of *layout* without first checking the conditions under which the *offset* is to be used.

Appearance is another objection to this type of *offset fitting*. When installed in exposed work, this *offset* does not look as good as the regular *ogee offset*.

However, this is definitely a faster *method* and many *shops* use it almost exclusively.

Offsets

STEP 1. Locate the various points in the usual manner. The amount of *offset*, setting apart the top and bottom openings as required, is indicated by w in Fig 6-1. The height is shown at x.

STEP 2. Choose a random radius, such as a in Fig. 6-1. Use a short radius (experience will enable you to judge the best amount). However, for the approximately 34" long offset, a 4", 5", or 6" radius can be used. It is not very practical to use too small a radius, as this makes the throat with practically no roll.

STEP 3. Having selected the desired radius, scribe in arcs for the heel and throat, Fig. 6-1.

STEP 4. Join the arcs with straight lines using a *straightedge* placed against the throat at one end and the heel at the other, as shown in Fig. 6-2.

STEP 5. Add the necessary 3/16" allowance required for the **easy edger.** Having already allowed for the lock and flange, the cheek should be cut out, notched, and marked.

STEP 6. Cut out a second cheek from this *pattern*, being certain to mark it on the other side, as a **right** and **left** are required.

STEP 7. With a flexible rule, the length of the wrappers is quickly determined.

This type of *offset* can be blocked out with the *shear list*. Although the wrappers cannot be definitely calculated before making the cheek, it is only necessary to allow a little more than the estimated stretchout. This surplus is easily trimmed at the bench.

The shaded area shown in Fig. 6-2 represents the straight section that develops within an *offset* which is laid out in this manner.

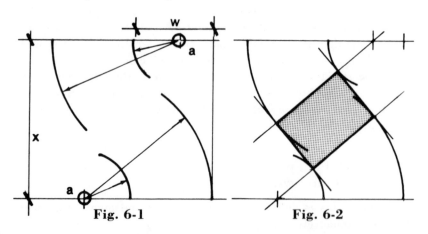

Fig. 6-1   Fig. 6-2

102

OFFSET No. 2
Regular Ogee, Method B

This type of **offset** is correctly proportioned; i.e., the radius is the required length to produce two identical angles that will be of the necessary degree to conform to the designated height, *b-x*, Fig. 6-3.

Comprised of both a concave and convex curved line which form an **ogee,** the *fitting* is referred to as an **ogee** *offset*. The *cutter* can, by choice, either develop the required radius point *j*, shown in Fig. 6-3, by applied mathematics such as described in **Math Problem No. 9** (see **Index**) or by using a prepared *chart* which is available for this specific purpose.

This *chart* gives the location of point *j* from the center point *e* as indicated by *Y* in Fig. 6-4.

Assuming that the *layout* is to be made in the regular manner, scribe a 7/8" line across the bottom of the sheet. At any convenient point on this line, which is usually about 1/2" from either end, erect a perpendicular line which will establish the point *x*. Draw a horizontal line parallel to the base line at the required height of the *offset*.

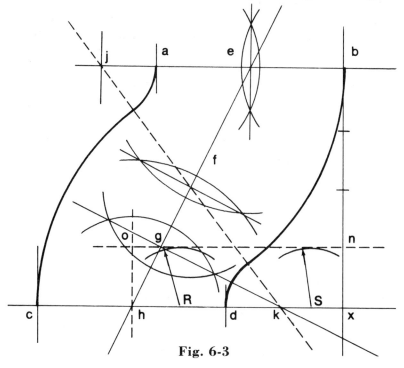

Fig. 6-3

Offsets

The length of both lines must be longer than given size of the opening of the *offset* and include the amount of offset desired. This will establish point *b* and, by stepping off the given amount of offset on the base line, this will then establish the point *d*, Fig. 6-3.

From both points, *b* and *d*, measure off the opening sizes of the *fitting* which will locate points *a* and *c*. When both opening sizes are the same, the following methods can be used to determine the radius point *k*.

Locate the center of both openings in the usual manner, points *e* and *h*. These points are connected by a straight line and the center point *f* of this line must also be located. Establish the center point *g*, which can be developed off the upper section, if desired. When using the *dividers* to determine points *f* and *g*, in particular, it is best to use wide intersecting arcs as shown. With a *straightedge* through these arcs, the radius point *k* is accurately located, Fig. 6-3.

NOTE: It should be made clear at this point, that Fig. 6-3 has been used to demonstrate a few different points regarding the layout. This accounts for the many lines shown which are not necessary in the actual development.

By dividing the height *b-x* into 4 parts and extending over parallel to the base, such as the dash line *n-o*, point *g* will be established, as shown in Fig. 6-3. This same condition would occur regardless of the angle of the line *e-h*. This is shown more clearly in Fig. 6-9. In Fig. 6-3, the center line, *o-h* is also 1/4 of the height and is used as the measuring point in order to extend the line from point *n* to establish point *g*.

However, the *short method* of establishing this same line is to apply the *Geometry Problem, Parallel line to a given one*, Fig. 12-20.

Determine 1/4 of the height, which can usually be done mentally, set the *dividers* to the size found, and scribe two short arcs, such as *R* and *S*, Fig. 6-3. Using a *straightedge* against these arcs, scribe in line *n-o*.

This relationship between the center line to the given height is based on the *Geometry Problem, Dividing a line equally*, Fig. 12-8.

The diagonal dash line through points *j* and *k* is not necessary in the actual *layout* of the *fitting*. It is shown only for the purpose of demonstrating how this line is the exact center of *f* and is the dividing point at which the curves meet in forming the outline of the *fitting*.

To complete the *fitting*, add the 3/16″ allowance for the **easy**

104

edger, and the allowance for lock and flange.

Using the cut-out cheek as a *pattern*, cut a second cheek which is turned over and marked on the other side, making a left and a right.

The length of the wrappers can be determined with a flex rule. A more efficient method is to calculate the length of the wrappers at the same time that the radius point is being figured by mathematics, as shown in **Math Problem No 9** (see **Index**) or by using the *offset chart*.

The shaded area of the cheek in Fig. 6-4 shows how the actual *offset* is divided into two identical angles when the *fitting* is correctly laid out.

NOTE: This **offset,** *Method No. 2, regular* **ogee,** *is also used as the basic principle in the development of the* **round offset,** *Method No. 2, shown in the Chapter on Round Layout Development (see* **Index***).*

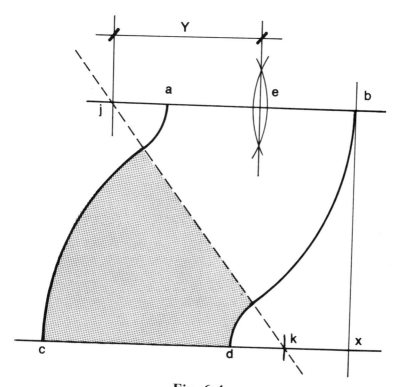

Fig. 6-4

Offsets

OFFSET No. 3
Unequal Opening, Method A

This type of **offset,** having two unequal openings, is developed here in two separate and different ways, Methods A and B.

Offset No. 3 is shown in its completed outline (the shaded area in Fig. 6-8) of the cheek which is correctly proportioned in relation to the given dimensions.

In developing this *offset* through this particular *method*, the *fitting* is worked up in much the same manner as the regular **ogee** *offset No. 2* of this section. Notice that each side of the *fitting* is developed as if it were an individual **ogee** *offset fitting*.

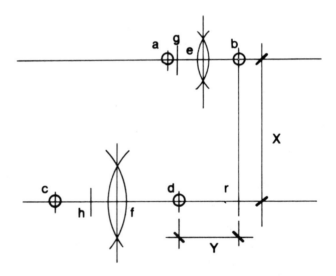

Fig. 6-5

STEP 1. Shear a piece of metal the height of X, as shown in Fig. 6-5, plus the required allowance for lock and flange. The length should be longer than actually necessary to include both the widest part of the *offset* opening and the amount of offset.

Scribe the edges at both the top and bottom of the sheet for the lock and flange.

STEP 2. In the usual manner, *lay out* various points of each opening in relation to the amount of required offset. This will then establish the points a, b, c, d and r, with the distance between d-r being the amount of required offset Y, Fig. 6-5.

STEP 3. Locate the center of each opening, *e* and *f*, but do not join with a connecting line as in the regular *ogee offset layout*.

STEP 4. The development of this *fitting* can be made by starting from either side. For this particular fitting the left side will be shown in its complete *layout* first.

RULE: Remember in this method of layout that the radius point will be established from the opening side where the center of the opening is found, and this will be the throat of the side being developed.

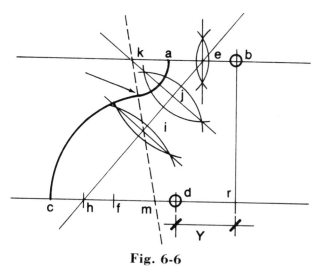

Fig. 6-6

STEP 5. Take the width *a-e* or *b-e*, which are both equal to 1/2 of the opening, and set this amount off from point *c*. This will establish the point *h*, Fig. 6-6. Then connect the points *e* and *h* with a straight line, as shown. This makes the left side of the *fitting* as a regular **ogee** offset and, as such, the actual procedure is handled throughout.

STEP 6. Locate the center *i*, then half of this again to find point *j*, in the usual manner. Square off from the line at this point and the radius point *k* is established, Fig. 6-6.

A line carried through the radius point *k* and the center point *i* will establish both the radius point *m* on the other end of the *fitting*, and the connecting point of the two arcs which form the left side of the *offset*. This line is shown as broken or dashed to avoid confusing the *reader*. It will be noted that the width which is clearly shown from points *k* to *b* and *c* to *m* is the same. Thus, it proves that, if the

Offsets

same arcs were to be placed from these same radius points *k* and *m*, with the throat and heel reversed, the fitting would definitely be a regular **ogee** *offset*, the same as shown in the previous *layout of* **Offset No. 2.** This fact is emphasized to show the *reader* the close similarity of the two *layouts*.

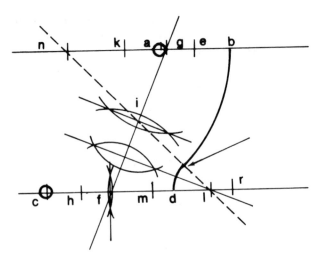

Fig. 6-7

STEP 7. The same process is now used in the development of the right side of the *fitting*, as shown in Fig. 6-7. However, the radius point is developed on the right side of the *fitting*, with the throat

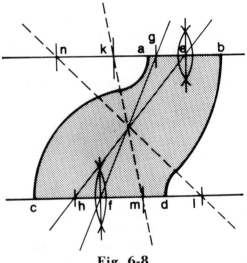

Fig. 6-8

being taken from the other end of the *fitting*. The radius is point *l* and the center point *i* remains the same. By carrying the dashed line through these points to the top, the radius point *n* is established for the heel. The line also notes the connecting point for the two arcs, in the same manner.

STEP 8. To complete the *fitting*, the required edges are added to the cheek in the usual manner. The second cheek is made as usual, as are the wrappers. The shaded area of Fig. 6-8 represents the complete *fitting* as it appears when developed in the manner just described. The various points and lines used in its construction have been included to show their relationship to the finished *offset*.

OFFSET No. 4
Unequal Opening, Method B

In order to show the actual layout of both **Offsets No. 3** and **No. 4,** the same proportions have been used in both cases. Although the Method A is theoretically the perfect *layout*, only a very slight variation occurs in B Method. Since this is, in the *author's* opinion, the better *method* to use, the difference should be overlooked.

The left side of the *layouts* will develop identically. However, the change occurs on the right side of the *fitting*, as the radius points do not develop in the same place. This causes the one side to be of a slightly different angle which will, of course, vary in accordance with the opening sizes of the *offset*. These, in turn, change the triangle, as shown in Fig. 6-10 by the shaded area.

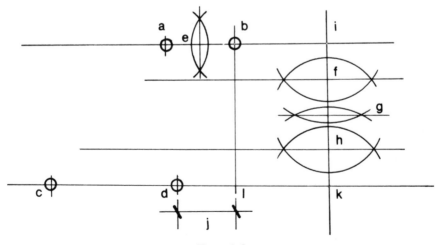

Fig. 6-9

Offsets

STEP 1. The preliminary development of the various points is handled in the same manner as described in **Offset No. 3,** establishing the various points as shown in Fig. 6-9. The opening points *a, b* and *c, d* with the width *j* as amount of offset between *d* and *l.*

STEP 2. The required height is between *i* and *k,* which must be divided equally into 4 parts. This can be done by finding the center point *g* and, from this and the outer points, making the intersecting arcs for *f* and *h.* By making the arcs larger than actually necessary, it is possible to obtain a much more accurate line through the *fitting.* In cases where a quarter of the height can be easily calculated, this amount can be measured off from the top and bottom, thereby eliminating the above process. It is possible to do this measuring directly at the center of the *fitting* to develop the points such as *r, s, t,* and *u,* as shown within Fig. 6-10. For another example of how this is best applied, see Fig. 6-3 *(ogee offset).*

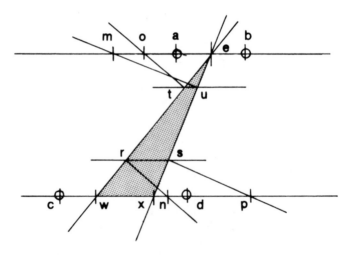

Fig. 6-10

STEP 3. The smaller opening of the *fitting* must be divided into equal parts which is in this case, the opening *a-b,* with point *e* being the exact center, Fig. 6-9.

One-half of this opening, or the width *a-e* or *e-b,* is set off from the points *c* and *d* of the larger opening at the other end of the *fitting.* These are shown as points *w* and *x* in Fig. 6-10.

STEP 4. Form the triangle *e, w, x,* as indicated by the shaded area of Fig. 6-10, with connecting lines. These lines establish the points *t, u, r,* and *s* as shown.

NOTE: *The triangles will seldom, if ever, be alike, since the amount of offset as well as the difference between the two opening sizes are the determining factors. However, once this is established, the actual development of the fitting will proceed as if it were two individual fittings.*

STEP 5. Using a *square* off the two lines of the *triangle* at each of the established points, *t*, *u*, *r*, and *s*, carry a line to the top and bottom of the *fitting* to locate radius points *m*, *o*, *n*, and *p*, Fig. 6-10. The two points which carry off the same line will be the radius points for the heel and throat for that particular side.

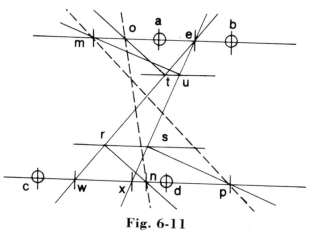

Fig. 6-11

NOTE: *Points m and p, in Fig. 6-11, are connected with a broken line to show that this line crosses at the exact connecting point of the two arcs as in Fig. 6-12 (see* **arrow**).

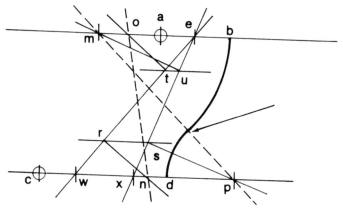

Fig. 6-12

111

Offsets

Also shown in Fig. 6-11 is the broken line opposite from points *o* and *n*, which in turn crosses the two arcs of this side also at the connecting point, shown in Fig. 6-13.

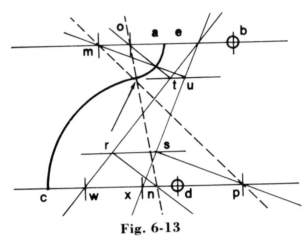

Fig. 6-13

STEP 6. Using points *o* and *n* for radius points, scribe arcs for the opposite side of *offset*.

The completed outline of the *offset* is shown in Fig. 6-14, with the shaded area indicating the actual shape of the cheek as it appears when correctly laid out.

STEP 7. To complete the cheek, add the required edges as necessary; then cut out and make another of the same, with the marking placed on the other side, making a *right* and *left* cheek.

STEP 8. Lay out the wrappers in the usual manner to complete the *fitting*.

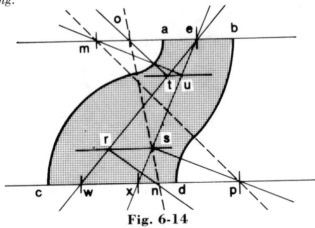

Fig. 6-14

OFFSET NO. 5:
Unequal Openings with Gores

This particular *layout* is used in the development of the large **offset** *fitting* which is too large to be made from one piece of metal. Thus, it becomes necessary to make it in **gores**.

The actual theory of development is the same as applied in **Offset No. 4**. Fig. 6-19 clearly shows the complete outlined *layout* of the *gored fitting* described here.

Although the completed *fittings* are the same, the actual development of the *gored fitting* must be applied in a different manner.

For this particular problem, assume that the height remains the same throughout. Although the same *method* of development is used for an *offset* having a rise or drop from one opening to the other, it is necessary to determine the actual length of each cheek before the *layout* is made.

This is done in the same manner as applied to the straight *transition fittings, in other words,* **triangulated.**

The *reader* is referred to the final portion of this section, described as **Offset No. 7,** the **transitional** or **compound offset.**

The development of this *fitting* is shown in outline, with the necessary allowances which are required for flanges, single-edges, etc., being omitted.

The drawing used here has not been laid out to scale, merely proportioned as considered best for explanatory purposes. However, various dimensions have been used to be certain the *reader* fully understands the developing process. With the various figures, each step may be checked thoroughly.

In developing this problem, the dimensions are applied as follows:

Overall height H =65" (given height)
Amount of offset O-c=40"
Opening a-b =30"
Opening c-d =38"

Using these given figures, the development of the **gored offset** is as follows:

STEP 1. Using a 4' sheet of metal, scribe a 7/8" line across the bottom. Locate the points b and a which represent the smaller opening of the *offset*, in this case 30", and then divide this in half to establish point j in Fig. 6-15.

Offsets

STEP 2. Knowing the given amount of the offset to be 40", add this to the 38" opening for a total of 78". Subtracting the 30" opening size of the other end makes the offset 48".

STEP 3. Draw a line parallel to the base line at a point equal to 1/4 of the finished height of the *fitting*. This is 16-1/4", since the overall height is 65".

Erect a perpendicular line at point *j* to cross the parallel line just placed, thereby establishing *z*, Fig. 6-15.

To the left of point *z*, on the horizontal line, step off 1/4 of each offset from point *z*, that is 10", to locate point *m*, and 12" to locate point *n*.

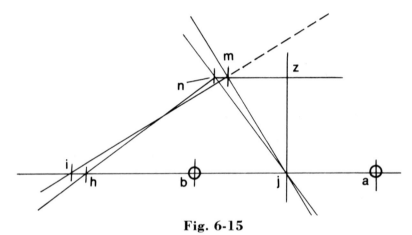

Fig. 6-15

STEP 4. Draw a straight line through points *m* and *j*; also, through points *n* and *j*, as shown in Fig. 6-15.

STEP 5. Placing a *square* at point *m* on line *m-j* and then at point *n* on line *n-j*. Carry 90° lines from these base lines to establish the points *i* and *h*. These are the radius points for both the heel and throat and, for this reason, it is most important that the *cutter* use care in their development.

STEP 6. As shown in Fig. 6-15, the line from points *i* to *m* is extended beyond the arc of the heel to *k* in Fig. 6-16. (Indicated by the broken line). This is 1/4 of the heel.

Also note that the line from *h* to *n* will cross the arc at 1/4 of the throat, as shown at point *g* in Fig. 6-16.

STEP 7. From the radius point *i*, swing the arc for the heel; from point *h*, swing the arc for the throat.

Portion A of the **offset**, at this time, represents only 1/4 of the *fitting*. Set the *trammel* points to the width b-g and use g as the radius point to swing a short arc to establish f. This also represents a half of this side of the completed **offset**.

STEP 8. In the same manner, using the width k-a and k as the radius point, locate l, as shown in Fig. 6-16. A straight line connecting points f and l completes section B which, added to section A, forms 1/2 of the completed *fitting* (Fig. 6-16).

STEP 9. Add 7/8" flange for the single edge of standing seam Y.

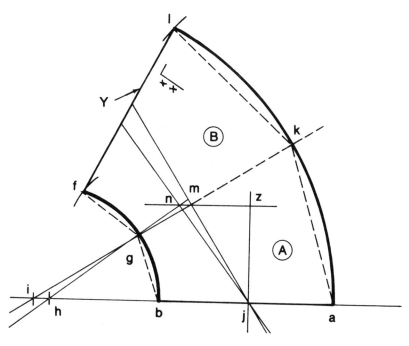

Fig. 6-16

STEP 10. In developing the second half of the **offset**, proceed in the following manner: Use a 48" wide sheet and scribe a 7/8" line across the bottom for the lock. Establish the 38" opening of the *fitting* on this line, as indicated by points c and d, Fig. 6-17.

Using the width equal to 1/2 of the smaller opening, step off from points c and d to locate points t and u. For this particular problem the width c-t and u-d will be 15", and the center span, t-u, 8".

STEP 11. Having determined the points t and u, the rest of the

Offsets

development is the same as described and shown for sections *A* and *B*.

Draw a line parallel to the base line at a point equal to 1/4 of the finished height of the *fitting*.

From points *t* and *u*, erect perpendicular lines to cross this horizontal line, thereby establishing points *p* and *s*, Fig. 6-17.

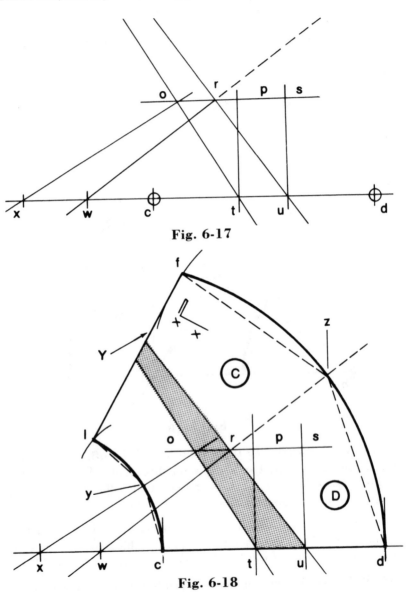

Fig. 6-17

Fig. 6-18

STEP 12. In the same manner as before, measure off 1/4 of each offset from the points *s* and *p*.

From *s*, measure off 12" to locate point *r* and, from *p*, measure off 10" to point *o*. A straight line is then carried through points *r-u* and *o-t*, as shown in Fig. 6-17.

STEP 13. Placing a *square* at points *o* and *r*, carry 90° lines across from lines *o-t* and *r-u* to establish the radius points *x* and *w*, Fig. 6-17.

STEP 14. Swing the arcs for the heel and throat, using point *w* as the radius point for the heel and *x* for the throat, as shown in Fig. 6-18.

Using *x* as radius point and with the *trammel points* set to the width *z-d*, swing a short arc to locate point *f*.

STEP 15. In the same manner, setting the *trammels* to the width *y-c* of the throat, locate point *l*. By connecting points *l* and *f* with a

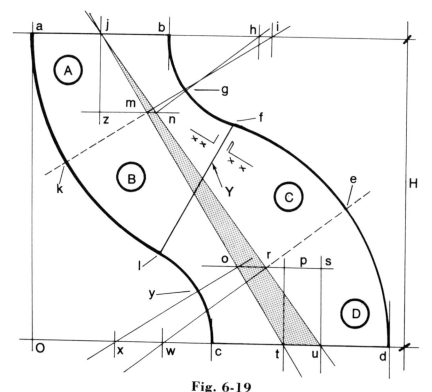

Fig. 6-19

117

straight line, the second half of the **offset** is completely outlined. Add the double part of the standing seam, Y, as shown in Fig. 6-18. Also add the required allowance to complete the *layout* for the cheek.

STEP 16. Make a right and left of both *gores*. A check of the *fitting* should show the span of *l-f* which is the standing seam Y, to be the same length in both *gores*. Also, the center lines, if correctly developed, should meet at this same line. This is clearly shown in the completed outline of the **offset**, Fig. 6-19.

STEP 17. The *wrappers* are developed in the usual manner and, if necessary, are made in two pieces.

NOTE: When making the gores with heavy material, it may be necessary to allow about 1/8" for growth at the standing seam. However, it is usually possible, when the wrappers are also made in 2 pieces, for the allowance to be omitted, as the growth should be the same in both cheeks as well as in the wrappers. The usual growth will be about 1/8" or less which is never enough to cause any difficulty.

NOTE: The dash lines, shown at the heel and throat of the gores in Fig. 6-16 and Fig. 1-18, are only used to call the reader's attention to the·quarter gores as they are developed. These lines have no specific part in the layout of this fitting and should be omitted in actual shop practice.

OFFSET No. 6
Irregular

Quite often the *cutter* will be required to make an **offset** of highly irregular proportions; e.g., the length of the **offset** is too short in comparison to the amount of *offset* that is required.

In order to keep within the given dimensions, the *cutter* must make an odd-looking, *blow-up* type of *fitting*, as shown in Fig. 6-20.

Although this *fitting* is *choked* across the span between points a and e, it is about the best that can be done under these circumstances.

It is sometimes possible to make the *offset* as shown in Fig. 6-21 in those cases that are not too extreme. Even then it will be most

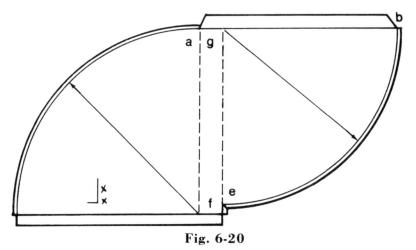

Fig. 6-20

difficult to keep the full capacity throughout the *fitting*.

Referring again to Fig. 6-20, notice that the radius, equal to the width of point *e* at the top of the *fitting*, is set off from point *b* to establish the radius point *g*. From *g*, swing the arc *e-b*.

The same procedure is applied in obtaining the radius point *f* for the other arc. The latter will, of course, be a little larger but this width is hardly sufficient to overcome the choked span from *a* to *e*.

In both *fittings*, point *e* is kept 1" from the flange to allow clearance for the *fitting* when placed into the government lock.

As shown in the *fitting* in Fig. 6-21, the radius is scribed from points *a* and *e*, the width of an opening such as *a-b* and *e-c*, respectively. The straight lines are then placed off these arcs, assuring the maximum width attainable.

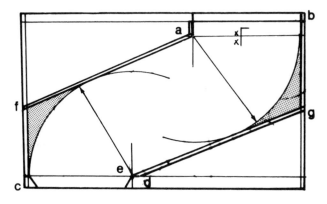

Fig. 6-21

Offsets

The actual development of this type of *fitting* must be kept within a limited space. However, the *cutter* can often take advantage of his own ingenuity in making the best of a bad situation.

The wrappers are made in the usual manner; however, in most cases, it will be necessary to measure off the stretchout with a flex rule. This is no problem.

The shaded area in Fig. 6-21 is only for the purpose of showing how the same *fitting* can be made without *cutting down* the area beyond the actual opening at *c-d*. Although this alternate may, to some extent, be a smoother flowing *fitting*, it has odd proportions. A better looking *fitting* results from leaving the squared-off effect as established by points *f* and *g*.

OFFSET No. 7
Transitional or Compound

Although each of the various **offset** *fittings* shown in this section has been described as being straight throughout, i.e., not offsetting up or down, this is not always the case. When the *offset* has to be raised or lowered at the same time, it is usually referred to as a *compound* or *transitional offset*. No specific *layout* has been shown for a *fitting* of this type, whereby either the top or bottom or both cheeks of the *offset* may drop or rise a given amount.

However, this can be required of any *offsets* that are shown here and, regardless of which *fitting* or the amount of change involved, it can still be handled in the manner described for each *offset*. It is only necessary to determine **true** length of the *offset* cheek, then follow through with the same method of *layout*, applied in the usual manner.

Usually, two separate layouts are required when the top and bottom cheeks are of a different amount of rise.

There are many *fittings* in which the difference between the two cheeks is very slight, possibly 1/4" or less. In such cases, use only one *layout* for both cheeks, then add this difference on the lock end of one cheek.

Some *cutters* use a *method* of *layout* similar to that used in the *layout* of the *transition elbow*. Dividing the cheek into a group of triangles and determining the **true** length, the new cheek is developed in the same manner. Theoretically, this is a correct *method* of *layout*. However, it is a time-consuming process, with no special advantages, making it most impractical for *shop* use.

120

chapter VII

ROUND LAYOUT DEVELOPMENT

INTRODUCTION

This *Chapter* explains the **layout methods** for various **round fittings** which the *cutter* in *air conditioning* work should be able to develop. Also included is a section describing the *method* of *procedure* to apply in developing a complete *layout* of *pipe* and *fittings* when working directly off a *plan* or from a *sketch* that has been sent into the *shop* from a *job*.

Although **round** work is generally considered to be the product of a *blow-pipe shop*, it is also used quite extensively in *commercial air conditioning installations*. The work may range from *boiler breeching*, made of 10 or 12 gauge *black iron*, to light gauge *galvanized iron fittings*, such as *elbows*, *tapers*, *tees*, etc.

Round fittings are not necessarily difficult to develop, but it cannot be denied that they are not handled as easily as *square* or *rectangular fittings*. To *lay out* a *round fitting*, particularly if it is made with heavy metal, and be assured that, when it has been formed, it will be the desired **shape** and **diameter,** requires a combination of **skill, knowledge,** and **experience.**

Usually, the *cutter* does not encounter **round layout** often enough to keep in practice, thus, when the occasion does arise, he is apt to approach it with considerable apprehension. However, whether the *cutter* has no knowledge of **round layout** or fully understands it but may need to *brush up* a bit, he will find the contents of this *Chapter* invaluable.

The *methods* of *layout* are shown and described as being applied directly on the metal, as required in actual *shop* practice. They are easy to apply, accurate, and dependable for developing any size *fittings* that may be encountered in *commercial air conditioning systems*.

ROUND ELBOW

Almost every *sheet metal worker*, at one time or another, has been shown how to *lay out* a **round elbow.** However, being able to *lay out* and *develop* the *elbow* from start to finish and produce a clean-

Round Layout Development

cut, accurately-made *fitting* offers a challenge to anyone, regardless of the extent of his experience.

Each step in the development of the *elbow* must be handled carefully, since it is quite possible for an error to be made in the *layout* process and not be detected until the *fitting* has been completed. The most critical part of the *elbow layout* is the development of the *miter line.*

Since this *line* will be determined by the amount of *miter rise*, the need for accuracy in establishing this figure cannot be emphasized too strongly.

It is the practice in most *shops* to have the *cutter* not only develop the *pattern* for the *elbow* but to follow through with the fabrication of the *fitting*, as well. It is, therefore, to the *cutter's* advantage that he develop a reliable *method* of *layout* which he can readily apply, in order to keep the *layout* time down to a minimum.

It is very difficult to make a close estimate of *how long it takes* to develop the *round elbow*. It may be possible to make up a certain *elbow* in about in hour; another one may take up to 2 or 3 hours. There are many things to be considered when attempting to make an estimate of time: e.g., the **diameter size,** the **number of gores,** the type of **seam** on the sides of the *gores*, the **method** of **joining** the *gores*, and — one of the most important items — the **gauge of metal** required.

When making just one *elbow*, the *cutter* is at a disadvantage in the time it takes him, especially when he also must develop the pattern. Any number of additional *elbows* can be made from the *pattern* and, when working up in quantities, it is possible to stack the metal and to use the *band saw* to cut out a large number of pieces at one time. This manner of handling multiple *fittings* is described also in *Chapter I, Cutting with a System*. It is always possible to make better time when developing several *elbows*, or any other type of *fitting*, of the same type and size than when making a single *fitting*. It will not necessarily show any change if *clocked* as an individual *fitting*. However, when the job is completed and the number of *fittings* made is checked against the total time taken, the average time per *fitting* will be considerably less.

The *elbow layout* is not difficult, if the basic fundamentals of its development are thoroughly understood and the *cutter* applies each step in the proper order. Three figures must be known before developing the elbow: they are the **diameter,** the **radius,** and **number** of **pieces** or **gores.**

NOTE: *When ordering out a* **round elbow** *on the work sheet, it can be described as having a certain number of* **pieces** *or* **gores;** *in either case, both terms refer to the number of sections which make up the finished elbow. When calculating the number of* **segments** *or* **pieces** *into which an elbow must be divided, in order to determine the degree of angle for the miter line, attention is called to the fact that the term* **pieces** *is applied here also but for a different purpose. This is explained in* **Math Problem No. 8.**

Knowing the *diameter* of the *elbow,* the *radius* required, and the number of *gores,* it is then possible to determine the amount of **miter rise** and follow through with the development of the curved **miter line,** as shown in Figs. 7-3 and 7-6. There are several *methods* of determining the amount of **miter rise** and it is only a matter of the *cutter* deciding which *method* he prefers. Either *method* is correct, since they are based on the amount of **rise** per **inch,** and the results will be the same.

Those *readers* who prefer using a prepared *Elbow Rise Chart* are referred to the *chart* shown in Fig. 7-1. This *chart* gives the amount of *rise* per *foot.*

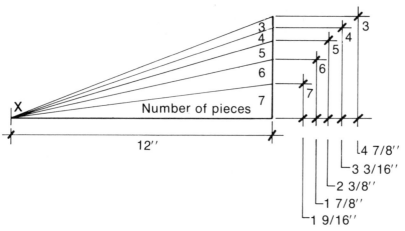

Fig. 7-1

To use this *chart,* proceed as follows: select a piece of metal much larger than the combined lengths of the *radius* and *diameter* of the *elbow* being developed. Place the metal on the bench and scribe a line lengthwise across the bottom of the sheet. Establish a starting point about an inch or so in from either end of the sheet; the intersection of this line and the one across the bottom will represent point *x,* as shown in the *Elbow Rise Chart,* Fig. 7-1.

Round Layout Development

From point *x*, measure off 12" and erect a perpendicular line upward from the base of the sheet. From the baseline, measure up the amount of *rise* indicated on the *Elbow Rise Chart*, which is shown opposite the number of *pieces* or *gores* in the *elbow* being developed. Measure this height off on the already drawn perpendicular line. This intersecting point will represent the amount of *rise per foot*, which is the same as shown in the *chart*. Place a *straightedge* diagonally on the sheet so that it connects both established points. With a *scratchawl*, draw a line from point *x*, on through the established height of the *miter rise per foot*, to beyond the length of the sum of both the *radius* and *diameter* of the *elbow*.

This diagonal line, at this particular point, has no specific connection with the *elbow* being developed, other than the fact that it is the correct *degree* of *angle* for the number of *gores* or *pieces* in the *elbow*. Any other *elbow*, regardless of its *diameter*, can be developed on the same line, providing it has the same number of *gores*. This fact is made quite clear in Fig. 7-2, which shows the effects of not applying the correct *radius* when using the *Elbow Rise Chart*. This sketch represents two 5-piece *elbows*, having the same amount of *rise*, and it shows what happens if the wrong *radius* is used. It is for this reason that extreme caution must be taken when attempting to use an old *elbow pattern*, when the radius is uncertain. Although it may well be the correct *diameter*, the same number of

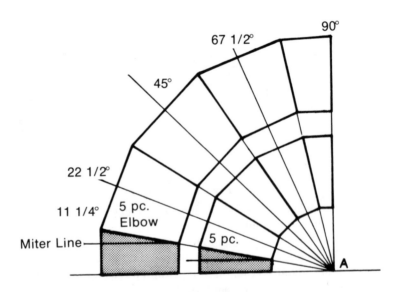

Fig. 7-2

pieces, and the correct amount of *rise*, it may **not** have the required *radius*. This is easily checked out but can be easily overlooked.

The established diagonal *miter line* represents the required *degree* of *angle* for one *segment* or the first end *gore* of the *elbow*. When the *radius* has been measured out from point *x*, erect a perpendicular line upward to the *miter line*. Then, from this line, measure off the *diameter* of the *elbow* being developed and also erect a perpendicular line up to the *miter line*. If all measurements are correct, this will then represent the side or elevation of the first *end piece* or *half-gore* (indicated by the shaded portion of Fig. 7-2).

The amount of **miter rise** is equal to the difference between the **heights** of the *throat* and *heel*, and this **rise** remains **constant** regardless of what the *radius* or *diameter* of the *elbow* might be. Although the *miter rise* does not vary, it is the *radius* which changes the height of the first *end piece* or *half-gore* and, consequently, the overall height of the finished *fitting*, as shown in Fig. 7-2.

As it is the *author's* policy to always try to apply *layout methods* which offer the least possible chance for error, it is his preference, when developing a *round elbow pattern*, to determine the *elbow rise* by the *method* shown in **Math Problem No. 4.** Both the amount of *rise* and the throat height are calculated by the same given **factor,** which is the same decimal figure equivalent to the **rise** per **inch.**

With this *method*, there is considerably less chance of making an error and it is not necessary to make any further calculations. It is, however, a good practice to check out the *heel* after having determined the *throat* height, as this makes it possible to double-check the *miter rise*. Using the same **factor,** multiply this by the combined amount of *radius* plus the *diameter* of the *elbow*. The resulting figure should be the same as the sum of the *throat* height and the *rise*, and any discrepancy between them indicates an error in the calculations. By rechecking, the error can be quickly located, and the correction made, before the error is carried any further into the development of the *miter line*.

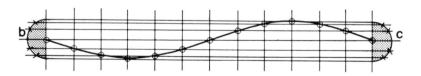

Fig. 7-3

Round Layout Development

Figs. 7-3 and 7-6 show two identical *miter lines*, each developed in a different manner. Although the *method* used in Fig. 7-3 is considerably faster and easier to apply on most *elbows*, the *method* described in Fig. 7-6 is more convenient to use on extremely large *elbows*.

Some *cutters* prefer to develop the *miter line* by means of either a half or quarter section *pattern*. This *method* works out very well and, again, it is just a matter of individual preference, as any of the *methods* is reliable. It is up to the *cutter* to select the *method* he is able to apply with accuracy and confidence. The same applies to determining the amount of **rise,** which has been discussed on preceding pages.

For explanatory purposes, it will be assumed that the *rise* is equal to the diameter of the half-circle, *d*, shown in Fig. 7-5. Also, the pipe diameter will be equal to the line *1'-7'*, which will, when extended upward beyond the half-circle *d*, establish the points of the *miter line E-F*, also shown in Fig. 7-5.

To follow through with this *layout method*, erect the half-circle shown in Fig. 7-4 and divide into equal parts, numbering from 1 to 7 inclusive. Extend these points downward through the *miter line* to the *base line*, thereby establishing the various *miter line* points, 1' to 7' inclusive. This completes the process for determining the various heights of the *base line* to the *miter line* in Fig. 7-5,

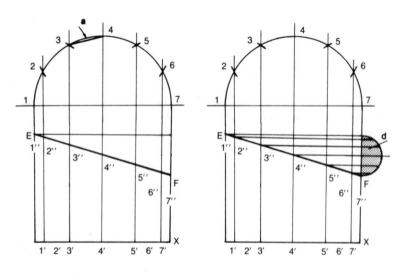

Fig. 7-4   Fig. 7-5

which, by transferring over to Fig. 7-6, will develop the progressive points of the *miter line* **curve.** When these points are connected with a continous line, this will represent the **end piece** or **half-gore,** since it is one-half of a center *gore.*

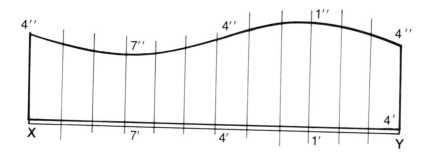

Fig. 7-6

The *base line X-Y* is equal to the *circumference* of the *elbow* being developed. This *stretchout* may be taken from a **Circumference Chart** (see *Chart Section*) or directly off the four-foot *circumference rule.* This *stretchout* will be correct in relation to the diameter selected and will be satisfactory to use for the lighter gauge metal, since one end of the *elbow* is usually run through the *crimper* to fit into the adjoining *fitting* or pipe of the same diameter. However, when heavy metal is used and an **inside** or **outside diameter** is required, this *stretchout* must be adjusted accordingly. This is explained in detail in the section, **Rolling Heavy Metal,** and also in **Round Development with Heavy Metal,** which is included in the last part of this *Chapter.*

Assume here that light gauge metal is used and the vertical side *seams* require one-inch allowance for the *acme* seam. This will be the blockout length, with the 48'' width of the sheet being used as it comes. Using a half-inch *scriber*, scribe lines across one long side of the *circumference* and the two 48'' ends. At the intersecting points of these three lines, place a prick mark, as these will be the starting points of the *layout.* Then, divide this *stretchout* into the same number of spaces or twice that of the half-circle of Fig. 7-4. Never use the width *a*, shown in Fig. 7-4, as a means of stepping out the divisions of the *circumference*, as this is actually the *chord* of that *arc*, and not the true length of the *arc* itself. Although this will, in most instances, be just a slight difference, when this difference is multiplied by 12 or 16 (according to the number of spaces used) it becomes a sizable amount which cannot be overlooked.

Round Layout Development

The best means of dividing the stretchout correctly is to divide the length in half and then into quarters, as it will be necessary to have four equal sections, regardless of whether each quarter is divided into three or four divisions. This is illustrated best in Fig. 7-7, where the actual *stretchout* extends along the base between points *a* and *b*, with the quarter sections being *c, d, e*. The *method* of dividing the quarters is shown, with *a-d* divided into four spaces and *e-b* divided into three spaces.

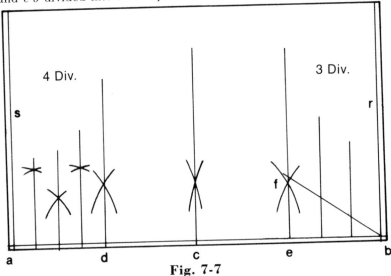

Fig. 7-7

The *dividers* are the best means of dividing the quarter section into four spaces; however, when the quarter is to be divided into three spaces, apply the *method* shown in the *Geometry Section*, Fig. 12-7, which describes *Dividing a Line of a Given Length*.

In Fig. 7-5, the various heights are given as 4'-4'', 5'-5'', etc. These are transferred to their respective places off the *base line of the pattern*, X-Y in Fig. 7-6, having carried up vertical or perpendicular *lines* from this *base line* prior to this step. With the various *points* having been established, the *miter line* is drawn in, to complete the *pattern* except for the necessary edge allowances.

The *cutter* must exercise extreme care in drawing curved *miter lines*, such as these; expecially in this case, as the *pattern* will be used to develop the center *gores*. This means that the *pattern* is turned over in order to establish both sides of the center *gore layout*. The slightest irregularity of the actual *miter line curve* of the *pattern* will show up when used in this manner.

It is important that this *miter line curve* be made as accurate as possible. No attempt should be made to hurry either the *layout* or cutting, especially with the latter, since the cut is usually made with a *unishear*, which must be guided carefully along the line.

In Fig. 7-3, this *method* is worked up in a different manner. A half-circle equal to the *rise*, as shown at *d*, Fig. 7-5, is placed at each end of a line equal to the length of the *circumference*, as shown at *b* and *c*, Fig. 7-3.

These half-circles are divided into an equal number of spaces. The *circumference* line also is divided equally, into twice the number of spaces as the half-circle. The points of these two half-circles are connected with lines drawn horizontally across the sheet. The points of the *circumference stretchout* line are carried across these lines vertically, thereby establishing the points of the *miter line curve.*

As shown in Fig. 7-3, start at the center of either end and work across the series of lines, connecting the successive intersecting points to obtain the *miter line curve.* The direct application of this

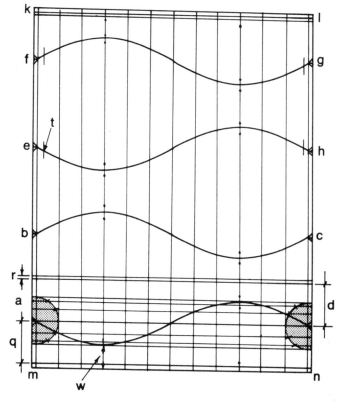

Fig. 7-8

Round Layout Development

method into actual *shop* practice is shown in Fig. 7-8, with the rectangular outline, *k*, *l*, *m*, *n*, representing the metal as it is blocked out at the *shear*.

The width *m-n* is equal to the required *circumference stretchout* of the *elbow*, plus a half-inch allowance on each side for the *acme* seam. The height is the width of the sheet, usually taken from 48″ wide stock. The procedure for obtaining the curved *miter line* is the same as that explained for the development of Fig. 7-3. The height, *q*, must be definitely established before starting the *layout*, as this height is critical to the development of the entire *fitting*. This dimension for *q* is determined by the radius of the *elbow* and should be calculated at the same time as the *miter rise*.

The layout shown in Fig. 7-8, since it is a hypothetical problem, has no specific dimensions. The various sections are proportioned as considered best for explanatory purposes only. The same applies to the *offset layout* shown in Figs. 7-12 and 7-13.

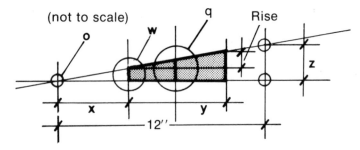

Fig. 7-9

The sketch shown in Fig. 7-9 demonstrates the relationship between all the heights and widths described throughout this section on the *layout* of the *round elbow*. The given *radius* is shown at *x* and the amount of *rise* in 12 inches is shown at *z*. Radius *x* determines the throat height *w*, which is shown in both Figs. 7-8 and 7-9.

By the same token, *x* will determine the *seam line q*, it being understood that the *diameter y* does establish *q*, but only in relation to the *fitting* itself, which is moved in either direction according to the desired *radius x*.

Referring again to Fig. 7-8, scribe a line across the length of the sheet, one-half inch up from the bottom and, again, at both ends. Develop the *miter line curve*, as already described, and carefully cut out this *end piece* or *half-gore*. Turn this around to match the contour of the proposed *center gore*. Scribe a line across the sheet,

130

against the straight side of the *half-gore* which is on top of the sheet, being certain that the curve of the *miter line* is even with the bottom curve. At this point, there are two *half-gores*, one being cut out and the other still on the sheet with only the curved *miter line* being cut.

The first cut *end piece* or *half-gore* should be kept for a *pattern* in case additional *elbows* are ordered. As this represents very little material, it is a good practice to do this with any *elbow*.

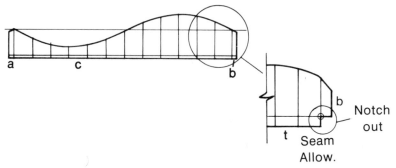

Fig. 7-10

The two corners of this *end piece pattern* should now be notched out, 1/2″ x 1/2″, Fig. 7-10. Turning the pattern over, place it against the straight line just drawn across the sheet. In this position, the half-inch of the *pattern* will be positioned in the same space as the half-inch on the bottom sheet, making the total allowance for the *center gore* one-half inch. This allows a quarter-inch on each *miter line* curve. Scribing a 1/4″ edge along the contour of the curves creates a suitable allowance for riveting the *gores* together. Use the vertical lines at their intersecting points as location for the rivets.

This *method* of using the lines is an accurate means of locating the holes for the rivets, as they will be exactly on the *miter line*. The two 1/4″ edges overlapping in this manner form a secure *seam*.

Scribe around the curve of the *miter line*, but do **not** cut out the *center gore* at this time. Prick mark the four points, *a*, *b*, *c* and *d* in Fig. 7-8, where the *miter* curve intersects the 1/2″ allowance scribed up the sides for the *acme* machine. With a pair of *dividers* set to the span between points *a* and *b* or *c* and *d*, which should be identical, step off the remainder of the sheet. This will establish points *e*, *f*, *g* and *h*. The *miter line* curve is now drawn in, as shown in Fig. 7-8, by placing the *pattern* at points *e-h* and *f-g*. Using a scratch-awl, scribe in the *miter line* curves to complete the *layout* of the *gores*. It is not always possible to obtain the full number of

Round Layout Development

pieces from one sheet, but additional material can be blocked out and developed in the same manner.

In cases where the top *half-gore* is not long enough to be used, it is a good idea to keep this piece for the *pattern* and use the original *pattern*, which was developed off the bottom of the sheet, as part of the *fitting*. However, be certain to mark this short *half-gore*, which is being kept for a *pattern*, with the actual height required at each end of the *pattern*, so that no mistake will be made if the *pattern* is used at a later date.

The *heel* and *throat* of each *gore* should be prick marked for the purpose of lining up the centers of the *gores* during the fabrication process. This should complete the actual *layout* of the *elbow*. Before putting this through the *acme* machine, notch in at each *cut* about 2", as indicated at *t*. This makes it easier to start and finish each cut through the *miter line* curve with the *unishears*.

The raw end or large end of the *elbow* will not require any additional metal; however, on the small end, which requires a *crimped* edge, it will be necessary to add about 1-1/4" more metal.

The 1/2" *seam* allowance, indicated at *r* in Fig. 7-8, should be sufficient for *seaming* the *gore*, either with *rivets* or by *spot welding*. Where very large *peened* joints are to be made, it usually requires a full 9/16", including 3/8" allowance for the double, and 3/16" for the single part of the *seam*.

For machine lock edges or adjustable *elbow* joints, the allowance might be anywhere from 1/8" to 5/16", depending upon the type of *elbow machine* available in the shop. As the gauge of metal used is also an important factor here, it is not possible to give a *set* figure for this type of *seam*, one that would apply to any *elbow* and still be correct in each case. Unless the *cutter* is aware of the allowance that is made for the particular machine he will be using, he should check on this with the *shop foreman*, rather than resort to the *trial and error* method which is a loss of time.

GROWTH ALLOWANCE

No special attention has been given here to *growth* allowance in the *elbow layout*. It has been the author's experience, in both *laying out* and *fabricating elbows* of different sizes and various gauges of metal, that no two *elbows* will develop in the same manner. Actually, it is only the smaller *radius elbows* which will have a tendency to grow in the throat, due to the difficulty in turning the edge in the *elbow* machine.

132

Two-piece Elbow

To overcome the problem, when making the *layout* for the small *elbows*, an allowance of one metal thickness can be made at the throat. This is made when the *miter line* is being established and should be set down towards the base of the sheet, from one metal thickness to nothing at the heel.

For the majority of the larger *elbows*, with a throat radius of ample proportions, there will be no problem and no allowance should be made. Most *shops* are inclined more towards *riveted* or *spot-welded seams* rather than using the *elbow machine*. When working up the *elbow* with these particular types of seams, care has to be taken as there is a tendency for the *elbow* to grow in the heel. This will develop the finished *fitting* with the heel being over 90° and, when checked with a *square* off the bench, the throat will not be in line with the heel.

It is possible to *take up* sufficient material at the *heel*, when using the *elbow machine*, to compensate for this expected growth. As a rule, 26 or 24 gauge *elbows*, having 4 or more *gores*, with a *radius* of 6" or more will not require any allowance.

Considering the numerous places throughout the *layout* and *fabrication* of the *elbow* where it is possible to encounter difficulty, it can be understood why the author claims each *elbow layout* is a challenge to one's ability.

TWO-PIECE ELBOW, 90 DEGREES
Short Method

A *two-piece elbow* is occasionally required for various purposes, and it is one of those *fittings* which should be made up as quickly as possible. The *method* of *layout* shown here is about the easiest and shortest possible means of development.

Block out the required piece of metal as indicated by letters *A, B, C, D* in Fig. 7-11. The length *A-B* is equal to the *circumference* of the *elbow*, which will be determined by its *diameter*, plus the 1" allowance for the *acme seam*.

The height *D-A* will be determined by the amount of *throat* specified, which is figured two times plus the required *diameter*, which is clearly shown in Fig. 7-11. Scribe a 1/2" line around the entire sheet, establishing the four corner points *a, b, c, d*.

Locate the center points *k* in the usual manner with the *trammels* from points *a-d* and *b-c* as shown. With the *trammels* or *dividers* set to the *radius* of the *elbow*, scribe the half-circles which will

Round Layout Development

locate points *t* and *s* on both ends of the sheet. Draw connecting lines from *t* to *t*, *K* to *K*, which establishes points *k* and *k*, and from *s* to *s*. With the *dividers*, divide each half-circle into quarters and carry lines across the sheet.

On the base line *d-c*, divide the *circumference* into eight parts, in the usual manner, and extend these points upward to the line *t-t* or above. This will establish the points for the *miter line* cut. Before making the cut, notch in about 2", as described for the *regular elbow layout*, and run through the *acme* machine.

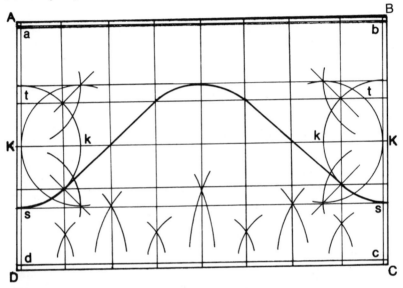

Fig. 7-11

ROUND OFFSET #1

By making a comparison of the *layout* for the **elbow** shown in Fig. 7-8 and the **offset** shown in Fig. 7-12, you can see that each of these *layouts* requires the same *method* of development, with the *offset layout* actually being derived from the *elbow pattern*.

Although making the *offset* in this manner is quite simple and is a definite time saver, it is not always practical since the *offset* has certain limitations.

Where the actual height of the *fitting* is of little or no importance, this *method* is most satisfactory. It can be compared to the similar *method* of using two 45° *ells* and a piece of pipe to make the *offset*.

134

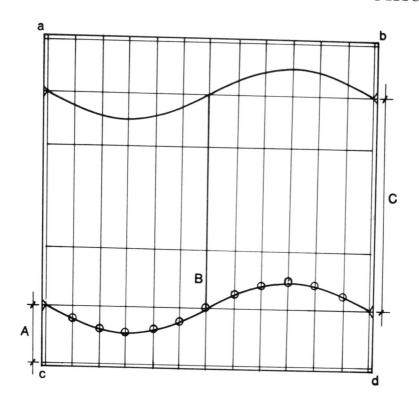

Fig. 7-12

The same situation exists whereby a definite height cannot be maintained and still produce the desired amount of offset.

When making this type of **offset,** in which a specified amount of offset and height must be maintained, it is necessary to make a *layout* using these given dimensions in order to develop the correct *angle* or *miter line.*

This requires a *side elevation* to be made of the *offset,* as shown in Fig. 7-13. This profile shows the entire *fitting;* however, this is not actually necessary, since the lower section *G* is required in order to establish the points on the *miter line.* Here, also, the *cutter* can apply either the small half-circle or the half-circle *G* in developing the series of points necessary for the *miter line.* The center *F* is made up of two corresponding *half-gores* or *end pieces* and a section of straight pipe, as indicated by the shaded area.

When the *miter line* is established, giving the height of the heel *Y* and the throat *X* shown in section *D*, it is a simple matter to use

Round Layout Development

Mathematics in determining the length of the center section *F*, Fig. 7-13.

Here, again, is a form of *triangulation* which is easily developed as follows: Assume the height as being equal to *W* in Fig. 7-13 and, having developed the *miter line*, both the throat *X* and the heel *Y* are also known. With the offset also established, it is only necessary to add the sum of *X* and *Y*, then subtract this from *W* to determine the height *N*.

This presents the **right triangle** *N*, *R*, *T* which may be described as: **base,** *N*; **height** or **offset,** *R*; and the **diagonal,** *T*, which represents the *hypotenuse.*

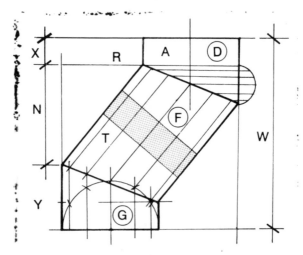

Fig. 7-13

If the line *T* were laid out in the drawing it could be measured directly, since this is a **true** length. However, when being calculated without being shown, it is obviously an unknown length.

By referring to **Math Problem No. 5,** which is shown elsewhere in this book (see **Index**), the *formula* for determining the **hypotenuse** can be found.

The rectangle shown in Fig. 7-12 represents the blocked-out metal from the *shear*, the width *a-b* being equal to the circumference of the *offset* plus 1" for the *acme* seam allowance. Height *a-c* equals the height that is determined by adding the two ends and the center section together, plus the allowance for making the seams and any possible edge that may be required. This entire *layout* is similar to the *elbow pattern* described in Fig. 7-8, and, therefore, can be handled in the same manner.

136

The height *A*, being the seam line, is equal to the center line shown in the profile, Fig. 7-13, also marked *A*.

The lines *B* and *C* are the same length as the rest of the lines throughout this center section. Therefore, using a *divider* or *trammel point*, the points of the lower *miter line* are placed directly in line, the length of *C*, and, when carried across the metal, a series of intersecting points will be made. These points are then tied together in a continuous curved *miter line* running parallel to the *miter line* on the bottom.

The required 1/2" allowance for the *mitered joints* can be added to the length *C* at the time the points are being transferred from the lower *miter line*.

The top section should also have an allowance for a probable *bead* and *crimp*, also 1/4" for the center connection. This can be taken care of by adding 1-1/2'' to the top section.

Notch and cut into the *miters* about 2" and run the *acme seams* through the machine before making the complete *miter* cuts.

On rather small *fittings*, it may possibly be better to hand-groove the seams, with allowance for this type of seam being three times the size of the finished seam.

ROUND OFFSET #2

This is another *method* of developing the **round offset,** such as shown in Fig. 7-13, when a specific **height** and **offset** is required. Although the actual development will be the same, once the *miter line* and length of center section is determined, this *method* offers a different approach in the manner of *layout* and absolute accuracy is assured without having to apply any *mathematics* other than the necessary given dimensions.

This *method* of *layout* is used in conjunction with the **Offset Chart** which the *cutter* uses for his *square* and *rectangular offset layouts*. From this **Chart,** a center-line *radius* point is given for a specific amount of *offset*, to be made within a definite *height*, developing a perfect *ogee offset*. For the purpose for which the **Chart** is to be used here, there is no need for the *stretchouts* which are included in the *Chart*.

So that this particular *method* of development will be thoroughly understood, a sample problem has been selected at random and is shown in its full outline in Fig. 7-14. This diagram is one of the few

Round Layout Development

sketches in the *book* that is set to *scale: being 1″* per foot. The purpose is to show the accuracy that can be obtained when the *round offset* is developed in this manner.

Although the complete *fitting*, in *elevation*, is shown in Fig.7-14, this is not necessary in actual *shop* practice. The various letters appearing in Fig. 7-14 are repeated in Fig. 7-15 so that they can be referred to in either sketch.

For this particular **offset,** the *fitting* is considered to be 30" high, 24" in diameter and offset 12". The center-line *radius*, shown on the **Chart** for such dimensions, is 21-3/4″

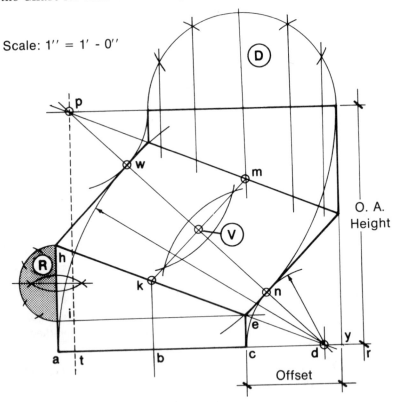

Scale: 1″ = 1′ - 0″

Fig. 7-14

. The following is a *step-by-step* procedure for developing the **miter line** and determining the length of center section required.

STEP 1. Place a sheet of *metal* on the *bench* and scribe a line across the bottom 1/2" up from the edge. Locate points *a* and *c*, which span the diameter of the fitting, in this case, 24″. Also locate point *b* and erect a perpendicular line part-way up the sheet, Fig. 7-15.

STEP 2. Refer to the *Offset Chart*, under *30"-High Offsets*, and locate 12" offset; this gives the center-line *radius* as 21-3/4". The stretchout is not required.

Measure out 21-3/4" from point *b* and establish *radius* point *d*. Also measure out 12" from point *c* and locate point *y*, Fig. 7-15. Set the *trammels* to the span of points *a-d* and swing an arc at a random length, as shown. With the *trammels* still set the same, from point *y*, locate point *t* on the same 1/2" line at the base. Setting the *trammels* to the span of *c-d*, also swing an arc of random length, Fig. 7-15.

STEP 3. Squaring up from the bottom of the sheet at point *t*, raise a perpendicular line in excess of the height of the *fitting*, which is 30". Setting the *trammels*, or by means of a *rule*, measure off 30" up from point *t* to establish point *p*. Using a *straightedge* held against points *p* and *d*, draw in the diagonal line shown, which will establish points *w* and *n*.

STEP 4. From points *w* and *n*, use the *trammels* in the usual manner to bisect the *line* and draw a line through the intersecting points of these arcs, which establishes point *v*. Scribe the arcs as far out as possible in order to obtain an accurate line through them. Also extend the line down enough to cross the perpendicular line which comes up from point *b*; the point of intersection is referred to as point *k*.

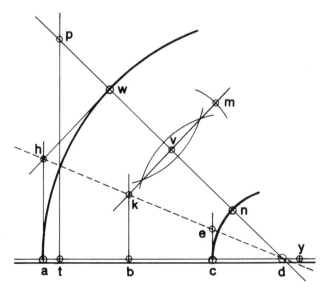

Fig. F-15

Round Layout Development

STEP 5. Using a *square* off line *p-d* at points *w* and *n*, carry lines down to locate points *h* and *e*. If all measurements are correct and care was taken, a *straightedge* placed across from points *h* and *d* should also cut exactly through points *k* and *e*. This diagonal line represents the **miter line.**

STEP 6. The span *k-v* represents 1/2 of the center section width. By carrying this width around to point *m*, from point *v*, you have the full length of the center section.

Having established the correct **miter line** and width of the center section, the **offset** is now handled in the usual manner. Referring to Fig. 7-14, either *method* of developing the rest of the *fitting*, as indicated at *R* and *D*, is up to the *cutter's* preference.

ROUND TAPERS

There are several *methods* of developing the **round taper** *fitting*. The choice of *layout* depends upon the **size** of the *taper*, **gauge** of **metal** required, and the extent of **accuracy** desired. This will be better understood after comparing the *taper layouts* that are shown here. The **taper** *fitting* is very simple to develop when the *fitting* is short and the dimensions are small. However, when the proportions become very large and the *fitting* is to be made from heavy gauge material, much more care must be taken in its development.

When the diameters are small and the height of the *taper* is fairly short, the *layout* is the same as used for either the **frustum of a cone,** as shown in **Math Problem No. 11,** or the *method* used for developing the **storm collar** (see **Index**). The best example is the **funnel** which includes both of these *layouts* in the development of the tapered *spout* and the top, cone-shaped *pouring section*.

The first type of **taper** *layout* described here is that which has large diameters and is of such length that the **frustum of a cone** *layout* is impractical. This is a **short method** for this particular *fitting*. However, like all *short methods*, it has certain limitations: in this case, that it can **only** be used if the *taper* is symmetrical. As this *method* is used by almost all *cutters* and since the symmetrical *taper* seems to be used more than off-center tapers, the *reader* should become thoroughly familar with its application. This *method* must be used with considerable care and discretion in order to obtain any degree of accuracy, particularly when the *taper* diameters are very large dimensions. Particular care must be used when developing the faired lines and when cutting out on these same lines.

Symmetrical Tapers

SHORT METHOD
Symmetrical, Any Diameter

STEP 1. Draw full a side view or *elevation* of the *taper, A, B, C, D*, Fig. 7-16.

STEP 2. With the *dividers* set at the width of *A-B* and with *A* as radius point, scribe the arc shown in Fig. 7-17. Using *B* as radius point, scribe another similar arc.

STEP 3. Using the width *C-D*, scribe arcs at the bottom of the *layout* from points *C* and *D*, Fig. 7-17.

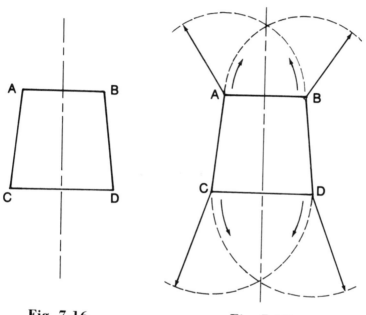

| Fig. 7-16 | Fig. 7-17 |

STEP 4. With the *dividers* set diagonally across the *layout* from points *A* to *D*, swing an intersecting arc from point *A* to establish point *C-1*, Fig. 7-18. With the *trammel points* still set the same, and using *B, C,* and *D* as the radius points, scribe intersecting arcs to establish the points *A-1, B-1,* and *D-1*, Fig. 7-18.

STEP 5. The *circumference* of both the top and bottom of the *taper* must now be determined. This is easily done, in the usual manner, using either the **Chart** (see **Index**) or the *bench rule*.

STEP 6. Either the top of bottom arc can be drawn in first. A *bench rule* or *flexible straightedge* should be curved into an arc

Round Layout Development

that touches all four points. Then draw in the arc, being certain to carry each line a little beyond each outside point, Figs. 7-19 and 7-20. This usually requires about three hands and it is best to have someone assist you in the process of drawing in the lines.

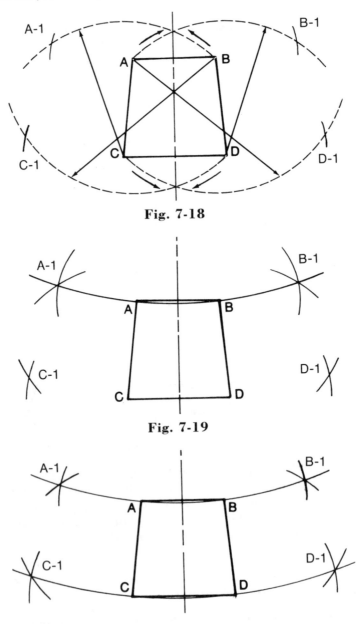

Fig. 7-18

Fig. 7-19

Fig. 7-20

Triangulated Tapers

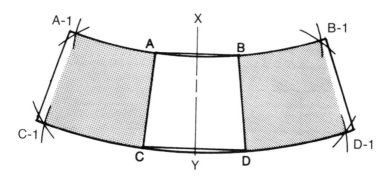

Fig. 7-21

STEP 7. From the center line *X-Y*, Fig. 7-21, using one-half of the top and bottom *circumference*, as required, measure out this length on one side of the center line only. Either side can be completed first, with the other side easily carried over with the *trammel points*.

STEP 8. To complete the *layout* of the *fitting*, add the required allowance at each side for the desired **seam**. This is usually made on the *acme* machine and requires 1/2" on each side.

These *taper fittings* almost always require *collars* and allowances must be made according to the type of *collar* desired. It is, in most cases, a good practice to use at least one gauge heavier metal than was used for the *fitting*. The *author* prefers to make the *collars* out of 20 gauge for the larger diameter *fittings*, in order to maintain the round shape.

*NOTE: Notice in Fig. 7-21 that a similar shape is outlined on each side of the taper being developed. This is **not** a part of the actual layout but is merely used here to demonstrate the purpose of developing the intersecting points, A-1, B-1, C-1, and D-1. When this method of development is applied, the actual layout is made by placing three elevation outlines of the taper side by side.*

ROUND TAPERS
Triangulation

This *method* of development can be applied to all *tapers*, regardless of **size** and the related position between the *openings*. This is undoubtedly the most reliable and accurate means of *layout* for any *taper*, especially for the *taper* whose size and proportions

143

Round Layout Development

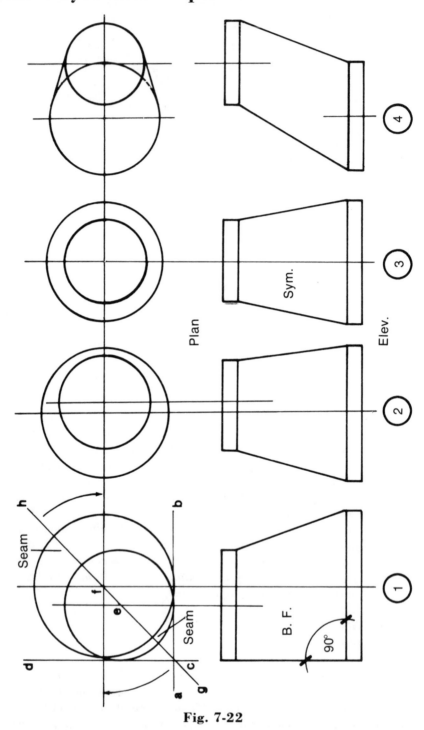

Fig. 7-22

144

make it impractical, and in most cases impossible, to be developed in the same manner used for *laying out* the *pattern* for *frustum* of a *cone*.

Although this is usually referred to as the *long method* of *layout* for the *taper fitting*, it only requires the development of a *half-pattern*, regardless of the position of the *taper*. Fig. 7-22 shows four different *tapers* as they appear for the *work sheet* in *elevation* and, directly above, as they appear in *plan view* which is the position required in order to develop them. The ability to develop a *pattern* for these four will make it possible for the *cutter* to handle any other *taper* he is required to make.

While it is easily seen that a *half-pattern* will be all that is required with *tapers #2, #3, and #4* this may not be apparent in regard to *taper #1*. Although this *taper* is assumed to be *flat* on the *bottom* and at 90° to the left-hand *side*, this *taper* will also only require a *half-pattern* in its development. This is due to the fact that, regardless of the position of the two openings of the *taper* in relation to each other, by locating the center of each and drawing a straight line across the *fitting*, it will automatically be split into equal half-sections and only require the development of a *half-pattern*.

Referring again to *taper #1*, Fig. 7-22, note that the centers of each opening, *e* and *f*, have a diagonal line, *g-h*, drawn directly through these centers. Also note how the straight side, *c-d*, and the flat bottom, *a-b*, position the *fitting*: however, by pivoting the *fitting* on the center *f* in the direction of the arrows, the *taper* will be placed in the same position as the others. The diagonal line will then rest upon the center line and the *fitting* will appear only slightly different from the others. The *seam* line will also be the same. All that is required with *taper #1* is to indicate, on the *pattern*, the exact center of both the *bottom* and the *left side*, and clearly mark the position in which the *fitting* is intended to be placed on the *job*.

The reader should note that *taper #3*, in Fig. 7-22, is the symmetrical *taper* that was developed in the *short method layout* which is also included in this section.

The *taper fitting* is quite versatile. It is able not only to change diameters between one end of the *fitting* and the other, but also to offset this *change* in any direction with considerable flexibility. Replacing a symmetrical *taper fitting* with one having a flat side offsets the line of *pipe* by a distance equal to one-half the difference between the two diameters.

145

Round Layout Development

In Fig. 7-22, *taper #4* illustrates how the *fitting* can be offset beyond its largest diameter.

In presenting the *method* of *layout* for *taper #1*, a larger drawing has been used in order to more clearly illustrate the detail which is so important in certain *layouts*.

This particular *taper* is one of that type and, for that reason, the layout is larger than the others shown in the pages of this book.

No difficulty of this nature is encountered in actual *shop* practice since the *fittings* are usually large enough to avoid any *spill over* of points needed in the layout. Such is not the case here and the *author* considers the development of this *taper* to be too important to the *reader* to chance having any part of its development passed over lightly.

TAPER #1
Bottom and One Side Flat

STEP 1. Select a piece of metal considerably larger than needed and scribe a line across the side and bottom of the sheet about an inch from the edge, *R,S,T,U*, Fig. 7-23. The intersection of these lines is point 2. From point 2, measure over the amount of the radius for the larger circle to establish point *a*. Set the *trammels* to this radius and, using *a* as *radius* point, scribe the half circle which will establish point *13*, with point 2 already being located. Locate point

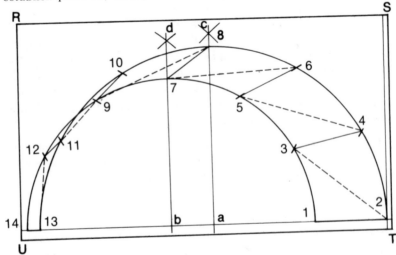

Fig. 7-23

146

c, shown in Fig. 7-23, with intersecting arcs from points *13* and *2*. With a straightedge, draw a line through points *a* and *c*.

STEP 2. With trammels still set to *radius* of the large circle, and from points *13, 8,* and *2,* divide each quarter into three parts, in the usual manner.

STEP 3. Determine the amount of offset between each center line to locate point *b* and follow through in the same manner as in *step 2* to develop the half-circle of the smaller opening. From points *14* and *1*, intersect the arcs as shown at *d* and erect perpendicular line *b-d.*

NOTE: The reason for developing the two sets of arcs is to be certain of obtaining a definite center line. One can never be certain by just squaring up from the bottom of the sheet.

 STEP 4. With the various points of each half-circle located, number and place the connecting lines as shown, using the dotted lines for the diagonals. These lines are of no constructive value to the *layout;* however, without them it would be quite difficult to keep track of each line as the *fitting* progresses towards the far side of the layout.

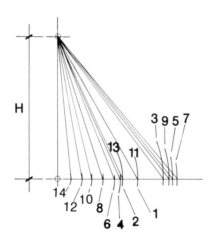

Fig. 7-24

STEP 5. Determine the true length of the various lines from the *layout* in Fig. 7-23, employing the usual *method* of *triangulation.* The finished height is shown at *H* in Fig. 7-24 and the various lines are handled in the same manner used to develop similar *fittings.*

STEP 6. The development of the half-pattern, Fig. 7-25, is handled

147

Round Layout Development

in the same manner as applied to *transition fittings.* Have two sets of *dividers,* with each set to one of the spans, such as *1-3* and *2-4,* and, by calculating the approximate length of half the greater circumference, *select a piece of metal for the half-pattern* of Fig. 7-25.

STEP 7. Determine the true length of line *1-2* and, with this as the

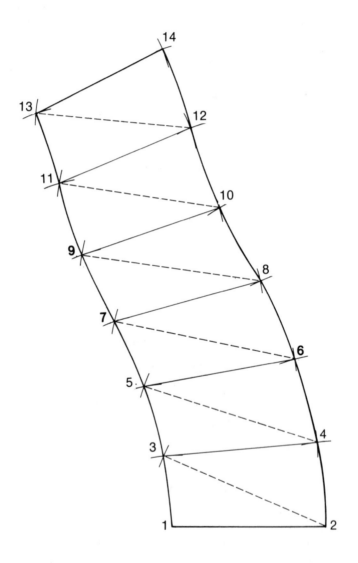

Fig. 7-25

starting point, proceed with the development of the *half-pattern* shown in Fig. 7-25. From this particulat point in the *layout*, the procedure is typical of other similar developments by *triangulation*.

STEP 8. With the series of points established and the curved lines drawn in position, Fig. 7-25, and before adding the allowance for the *acme seam*, measure out each stretchout to be certain of having one-half of the *circumference*. Allowance for the *collars* should also be made in the usual manner.

TAPER #3
Symmetrical

The development of **taper #3** is shown here in Fig. 7-26, with the *layout* being made from *1/4-pattern* which is all that is required. This *layout method* can be handled with considerable speed and, compared with the *short method* shown in this section for the same *taper*, there is very little difference in actual *layout* time.

No description is required here, as Fig. 7-26 is self-expanatory. Line *a-b* is the finished height and is used to *triangulate* the two lines, as shown.

The application is the same as for any *taper* developed in this manner and, for this reason, both *tapers #2* and *#4* are not shown here.

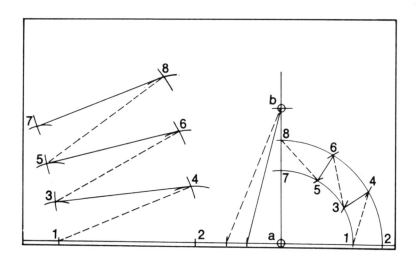

Fig. 7-26

Round Layout Development

SHORT TAPERS WITH COLLARS

The *cutter* is often required to make these short **taper fittings** since they are frequently used in high-pressure *air conditioning systems* employing *round pipe* and *flex hose*.

With this system, the *conditioned air* is carried to a *plenum* box. Then, by means of round *tapin collars* and *flex hose*, the air is carried to the various room outlets. The *outlet* opening size often runs several inches larger in diameter than the connecting *flex hose*. It is here that the *short taper* is required: to make this connection between the *flex hose* and *air diffuser*.

Fig. 7-27 shows the *short taper* which can have a finished overall height, *a*, of about 6 inches. Allowing about a 3/4'' edge on both top and bottom, the tapered section, *b*, is 4-1/2''+.

Fig. 7-27 also shows that, due to the short height of the *fitting*, measured to point *h*, the *layout* can be made in the same manner as that for a funnel.

Using the apex, *h*, as point of radius, the arcs can be carried around as required. Then, by using the circumference of either *d* or *j*, as shown in Fig. 7-29, measure off 1/2'' on both sides of the center line, in the usual manner of a cap or funnel layout. Use a spot-welded seam if possible.

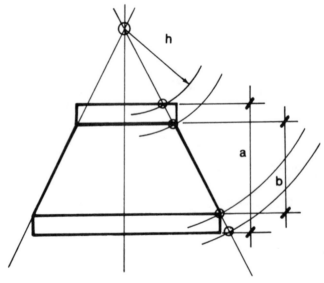

Fig. 7-27

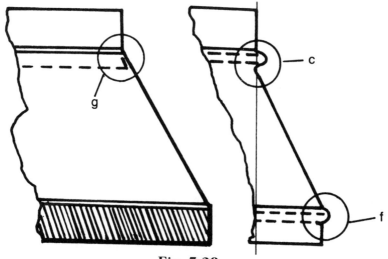

Fig. 7-28

If a separate *collar* is necessary, lap as shown at *g* and spotweld, rather than use the regular *collar* lock as shown in Fig. 7-28 at *c* and *f*.

Since the overall finished height, *e*, is seldom critical, finished openings *d* and *j* should be shaped by using the *crimper* to bring the *collars* up to the desired size. In this case, a separate *collar* can be used, as previously mentioned.

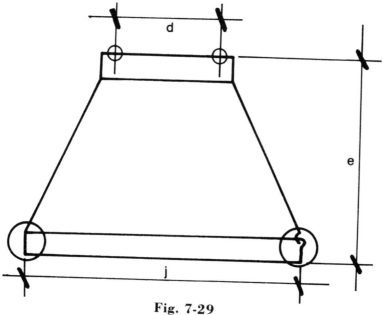

Fig. 7-29

Round Layout Development

LAYOUT OF TEES

There are seven types of tees that the cutter should be able to lay out. They are:

TEE NO. 1. A tee or pipe intersecting another pipe of a **Different Diameter, on Center, at 90°**

TEE NO. 2. A tee or pipe intersecting another pipe of the **Same Diameter, at 90°**

TEE NO. 3. A tee or pipe intersecting another pipe of a **Different Diameter, on a Tangent, at 90°**

TEE NO. 4. A tee or pipe intersecting another pipe of the **Same Diameter, at an Angle**

TEE NO. 5. A tee or pipe intersecting another pipe of a **Different Diameter, at an Angle**

TEE NO. 6. A tee or pipe intersecting another pipe of a **Different Diameter, on a Tangent, at an Angle**

TEE NO. 7. A tee or pipe intersecting a **Symmetrical Taper, at an Angle**

The *methods* of *layout* shown here are those used by the *author* himself and have proven to be most reliable. The fact that the *layout* is done directly on the material cut for the **tee** saves a considerable amount of time, since this can be figured out as the shear list is being prepared and can, therefore, be included.

The *layout* procedure, as used in each of the seven **tees**, is quite different in many ways and has a definite advantage over other *methods*. All of the *layout* work has been prepared so that it can be applied directly at the bench under actual *shop* conditions. Therefore, the described step-by-step procedure is explained with the regular tools being used in making the *layout*. In addition to the *square*, a metal *T-square* is invaluable, particularly in this type of layout work.

In several of the following pages you will notice that an extra line or quarter-circle has been placed in the *layout*, when, actually, only one would serve the purpose. This is true if the *layout* is made on a drawing board. However, working directly on the metal, using a *square* and *straightedge*, often requires two working points. A very good example of this is shown in the *layout* of **Tee No. 7,** Fig. 7-46.

Attention is called to the fact that in each of these *layouts* the *miter lines* are developed from the opposite side from that used in other *methods* of *layout*.

Layout of Tees

Although only the one *layout* is required for each of the seven **tees**, notice that two or more drawings have been made for each **tee.** The only purpose in doing this is to avoid any possible confusion regarding the procedure and explanation of the *layout method* being described. Because the drawings must be reduced down to such a small *scale*, the various points appear close together which, in many cases, gives the drawings the appearance of being complex and difficult to make. However, once the *layout* is understood, just the opposite proves to be true.

The half-inch edge scribed around the sheet allows for the seam on the ends and the top and bottom are used in the development of the *tee.* By working off the top line, the *tee* can be cut out directly on the *miter line;* then, using a scriber along this edge of a half-inch will mark off the lap allowance. This eliminates having to draw in the *miter line* twice as so many *methods* usually require.

NOTE: The layout of holes required for the various types of **tees** *shown here has been purposely omitted. In actual shop practice the cutter must keep his* **layout time** *to a minimum. It is, therefore, the practice of most cutters to make the* **tee** *and then, holding it in place, scribe around it to establish the location of the hole.*

However, attention is called to **Tee No. 2;** for, whenever a **tee** intersects a *pipe* of the same diameter, the development of the **tee** itself also forms a *half-pattern* for the hole *layout.* This is shown in Fig. 7-33, where the center portion indicated by *1** to *1** is the center of the *half-pattern.* By transferring down the points *2*, 3*, 4*, 3** and *2** to below this center line *1*-1**, the hole *pattern* is completed.

In actual practice, this can be applied more effectively by completing the **tee** *layout* and cutting it out in the usual manner. Then block out the *pipe* into which this **tee** is being made to fit. Establish a center line across the metal horizontally and also vertically. Using the developed **tee** as a *half-pattern,* place it upon the center line of the *pipe* so that points *1*-1** lie directly on the horizontal center line and, at the same time, point *4** rests upon the vertical center line. Scribing a line against the curve of the **tee** *pattern* will form one-half of the hole cutout. Then turn the *pattern* over and use the same procedure to complete the lower half of the hole cutout.

It is a good practice to never cut these openings directly on the line developed for the *hole.* While this will be the correct *hole lay-*

Round Layout Development

out and will match the **tee**, it is best to leave about 1/4″ or 3/8″ of stock on the *pipe*, and particularly in the corners at *1*-1**. After the **tee** is fitted into position on the *pipe*, this left-on material should be *dressed* down tight against the *pipe* itself. This makes a neat job which will not only look good but will be a perfect fit. Because it is often impossible to use a *hammer*, a piece of 3/4″ diameter *pipe* makes the best tool possible for this purpose. Not only is the pipe small enough to have swinging room inside the *fitting*, but it carries enought weight to make it highly effective.

TEE NO. 1. A Tee or Pipe Intersecting Another Pipe of a
Different Diameter, on Center, at 90°

STEP 1. Let the rectangle *A,B,C,D*, shown in Fig. 7-30, represent the blocked-out metal required for the **tee.**

The length *A-B* equals the circumference, plus the seam allowance. The height *A-C* equals the longest point of the **tee**, plus 2″ or 3″ working area.

STEP 2. Scribe a 1/2″ line completely around the sheet shown in Fig. 7-30. The lines on each end represent the seam allowance, with the top and bottom being used for *layout* purposes.

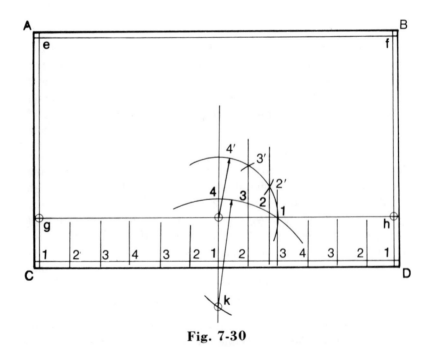

Fig. 7-30

154

STEP 3. Divide the **base line** into 12 equal spaces; number as shown. At the center of the base line or point *1*, use a *square or metal T-square* off the bottom of the sheet to erect a perpendicular line approximately half-way up the sheet.

STEP 4. From points *e* and *f*, at the top of the sheet, measure down the seam lines, the longest length of the **tee,** to establish points *g* and *h* in Fig. 7-30.

Using a *straightedge* from these points, draw a line across the sheet, locating point *j*.

STEP 5. With the *dividers* set to the radius of the **tee** and using *j* as the radius point, scribe the quarter-circle, establishing points *4'* and *1*, Fig. 7-30.

With the same radius, scribe the intersecting arcs from points *4'* and *1* to locate the points *3'* and *2'*.

STEP 6. Point *k* must be located off the sheet. Usually, by just pushing the sheet upwards, the bench will serve the purpose.

Extend the center line below the sheet and set the *dividers* to the radius of the intersected pipe. From radius point *1*, cross the center line with a short arc, thereby establishing the radius point *k*, Fig. 7-30. Using the same radius from point *k*, scribe the long arc through point *1*, locating point *4* where it crosses the center line.

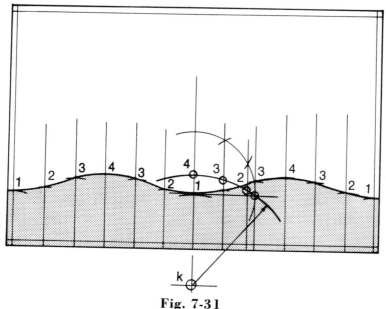

Fig. 7-31

Round Layout Development

STEP 7. Using a *square* off the bottom of the sheet, extend the points *3'* and *2'* downward to locate points *3* and *2* on the large arc, which represents a portion of the intersected pipe.

STEP 8. Extend the points of the base line upwards, using either a *T-square* or *square* off the bottom.

By means of the *dividers*, carry these points of the large arc to the lines of corresponding number, Fig. 7-31.

STEP 9. Carefully connect the points to form a continuous *miter line*. Cut out directly on the *miter line* and scribe a 1/2'' line along this edge for the required lap. This completes the *layout*.

TEE NO. 2. Tee or Pipe Intersecting Another Pipe of the **Same Diameter, at 90°.**

STEP 1. Let the rectangle *A, B, C, D*, shown in Fig. 7-32, represent the blocked-out metal required for the **tee.**

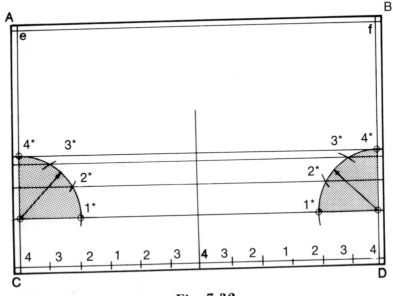

Fig. 7-32

The length *A-B* equals the *circumference*, plus the seam allowance. The height *A-C* equals the longest point of the **tee**, plus 2'' or 3'' working area.

STEP 2. Scribe a 1/2'' line completely around the sheet shown in Fig. 7-32. The lines on each end represent seam allowance, with the top and bottom being used for *layout* purposes.

STEP 3. Divide the **base line** into 12 equal spaces; number as shown.

STEP 4. From points *e* and *f*, at the top of the sheet, measure down the seam lines the longest length of the **tee**. This establishes the points *g* and *h*, Fig. 7-32. Using a *straightedge* from these points, draw a line across the sheet.

STEP 5. From points *g* and *h*, using the radius of the **tee**, scribe a quarter-circle which locates points *4** and *1**.

With the same radius, scribe the intersecting arcs from points *4** and *1** to locate the points *3** and *2**. Connect these points across the sheet, as shown in Fig. 7-32.

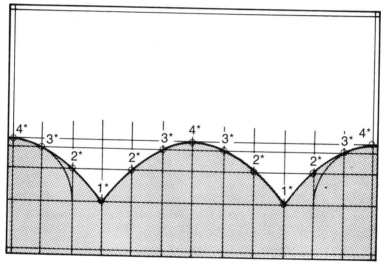

Fig. 7-33

STEP 6. Using a *square* from the bottom of the sheet, extend lines upward from the points on the baseline, as shown in Fig. 7-33, to intersect the horizontal lines already on the sheet. These points of intersection establish the **miter line** for the **tee**.

STEP 7. Carefully connect the points to form a continuous, smooth *miter line*. Cut out directly on this *miter line* and scribe a 1/2″ line along this edge for the required lap, thus completing the **layout**.

TEE NO. 3. A Tee or Pipe Intersecting Another Pipe of a **Different Diameter, on a Tangent, at 90°.**

STEP 1. Let the rectangle *A,B,C,D*, shown in Fig. 7-34, represent the blocked-out metal for the **tee**.

Round Layout Development

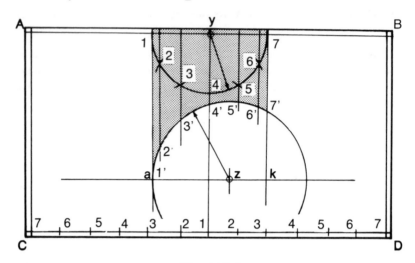

Fig. 7-34

The length *A-B* equals the circumference, plus the seam allowance. The height *A-C* equals the longest point of the **tee**, plus 2'' or 3'' working area.

STEP 2. Scribe a 1/2'' line completely around the sheet shown in Fig. 7-34. The lines on each end represent seam allowance, with the top and bottom being used for *layout* purposes.

STEP 3. Divide the **base line** into 12 equal spaces; number as shown. At the center of the base line or point *1*, use a *square* or metal *T-square* off the bottom of the sheet to erect a perpendicular line to the top of the sheet, locating point *y*.

STEP 4. With the *dividers* set to the radius of the **tee** and using *y* as the radius point, scribe a half-circle which locates points *1*, *4*, and *7*. Then in the usual manner determine the points *2*, *3*, *5*, and *6*, Fig. 7-34.

STEP 5. Using a *square* off the bottom of the sheet, extend the points *1* to *7*, from the half-circle down to the bottom base line, as shown in Fig. 7-35.

STEP 6. With the *dividers* set to the longest side of the **tee** and from points *1* and *7*, scribe short arcs that intersect the vertical lines from points *1* and *7*, and establish the points *a* and *k*. Using a straightedge placed along points *a* and *k*, carry a line through these points, as shown in Fig. 7-34.

STEP 7. Setting *dividers* at the radius of the intersected *pipe*, from point *a* scribe a short arc to locate point *z*. Using the same radius

158

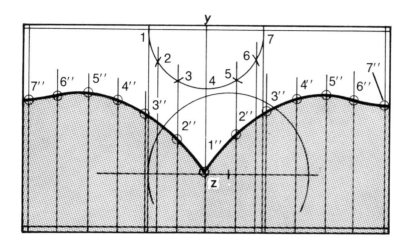

Fig. 7-35

from point *z*, scribe an arc long enough to intersect all the extended lines from the **tee** in order to establish the points *2′, 3′, 4′, 5′, 6′*, and *7′*, as shown in Fig. 7-34.

STEP 8. Using a *square* off the bottom of the sheet, extend the points of the base line upward, to slightly beyond the proposed *miter line* that will be developed between the highest point at *1′* and the lowest point *5′*.

STEP 9. Using the *trammel* points from the base line up to points *1′* through *7′* inclusive, carry each length over to its respective place on the numbered line; this will develop the contour of the *miter line*. The corresponding numbers are indicated by *1″, 2″, 3″*, etc., Fig. 7-35, on the *miter line*.

STEP 10. Carefully connect the points to form a continuous, smooth *miter line*. Cut out directly on this *miter line* and scribe a 1/2″ line along this edge for lap. This completes the *layout*.

The shaded area in Fig. 7-34 shows the **tee** in profile, and clearly illustrates how the **tee** sets on the larger *pipe*.

TEE NO. 4. A Tee or Pipe Intersecting Another Pipe of the
Same Diameter, at an Angle.

STEP 1. Let the rectangle *A, B, C, D*, shown in Fig. 7-36, represent the blocked-out metal required for the **tee**.

The length *A-B* equals the circumference, plus the seam allow-

Round Layout Development

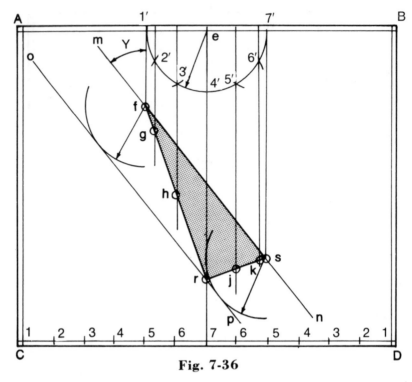

Fig. 7-36

ance. The height *A-C* equals the longest point of the **tee**, plus 2″ or 3″ working area.

STEP 2. Scribe a 1/2″ line completely around the sheet. The lines on each end represent the seam allowance, with the top and bottom being used for *layout* purposes.

STEP 3. Divide the **base line** into 12 equal spaces; number as shown in Fig. 7-36. At the center of the base line, or point 7, use a *square* or metal *T-square* off the bottom of the sheet to erect a perpendicular line to the top of the sheet, locating point *e*.

STEP 4. With the *dividers* set to the radius of the **tee** and using *e* as the radius point, scribe a half-circle which locates points *1′*, *4′*, and *7′*. Then, in the usual manner, determine the points *2′*, *3′*, *5′*, and *6′*, Fig. 7-36. Squaring up from the base, extend lines vertically to the bottom of sheet from all seven points of the half-circle.

STEP 5. Measure down from point *1′* using the desired length of the throat, to establish point *f*.

From this point place a diagonal line *m-n* using the required degree of angle, as indicated at *Y*. Line *m-n* represents the outside of the intersected pipe and also locates point *s*, as shown in Fig. 7-36.

STEP 6. Using the same radius as the **tee** and with points *f* and *s* as radius points, scribe two short arcs as shown in Fig. 7-36, thereby establishing the parallel center line *o-p*, and also point *r*. If the *pattern* is correctly placed, the corner at *r* will form 90°.

STEP 7. Using straight connecting lines from points *f* to *r* to *s*, establish the *miter line* between the two intersecting pipes, and locate the points *g*, *h*, *j*, and *k*, Fig. 7-36.

NOTE: Whenever pipes of the same diameter intersect each other, the miter line is connected with straight lines.

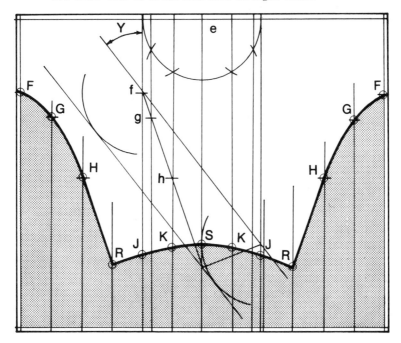

Fig. 7-37

STEP 8. Using a *square* off the bottom of the sheet, extend the points of the base line upward, as shown in Fig. 7-37. From the same base line, using *dividers* or *trammel* points, span the distance to the points *f* through *r* inclusive, transferring each point to its respective line, thereby establishing these same points on the *miter line*. The corresponding points are marked with the same letters, but shown in **capital** letters.

STEP 9. Carefully connect these points to form a continuous, smooth *miter line* and cut out on this same line. Scribe a 1/2″ line

Round Layout Development

along the *miter line* for the required lap, completing the *layout*.

TEE NO. 5. A Tee or Pipe Intersecting Another Pipe of a
Different Diameter, at an Angle.

STEP 1. Let the rectangle *A, B, C, D*, shown in Fig. 7-38, represent the blocked-out metal required for the **tee**.

The length *A-B* equals the circumference, plus the seam allowance. The height *A-C* equals the longest point of the **tee**, plus 2″ or 3″ working area.

STEP 2. Scribe a 1/2″ line completely around the sheet. The lines on each end represent seam allowance, with the top and bottom being used for *layout* purposes.

STEP 3. Divide the **base line** into 12 equal spaces; number as shown in Fig. 7-38. At the center of the base line, or point 7, use a *square* or *T-square* off the bottom of the sheet to erect a perpendicular line to the top of the sheet, locating point *e*.

STEP 4. With the *dividers* set to the radius of the **tee** and using *e* as the radius point, scribe a half-circle which locates points *1′*, *4′*, and *7′*. Then, in the usual manner, determine the points *2′*, *3′*, *5′*, and *6′*, Fig. 7-38.

STEP 5. Using a *square* off the bottom of the sheet, extend the points *1′* to *7′* downward; at the same time, extend points *5′* and *6′* up to the top of the sheet, locating points *i* and *j*, Fig. 7-38.

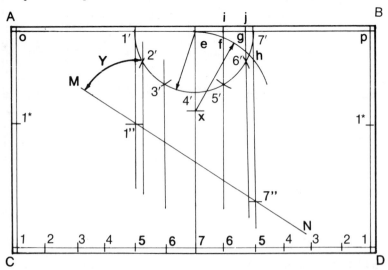

Fig. 7-38

162

STEP 6. With the *dividers* set to the radius of the larger intersected pipe, from point *e* on the center line, establish the radius point *x*. Using the same radius from point *x*, scribe the arc as shown from point *e*, thereby establishing points *f*, *g*, and *h*.

NOTE: This part represents the **tee** *intersecting the pipe and is a view rotated 90° from portion below diagonal line M-N.*

STEP 7. Measure down from point *1'*, using the desired length, to locate point *1''*. From this point draw a diagonal line *M-N*, using the required degree of angle, as indicated at *Y*, Fig. 7-38. This also establishes the point *7''* on the long side of the **tee**.

STEP 8. The line *1'-1''*, shown in the throat, is actually the seam line and will remain the same length. Therefore, it can be transferred to the outside seam lines, as shown in Fig. 7-38, indicated by points *o-1** and *p-1**.

STEP 9. Square out from the diagonal line, *M-N*, at points *1''* and *7''* and draw in a short line as shown in Fig. 7-39.

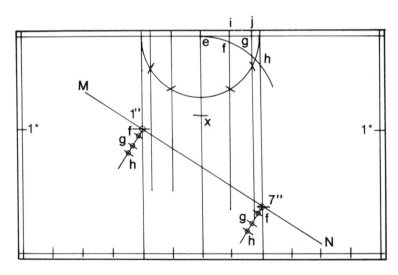

Fig. 7-39

STEP 10. From these same two points, *1''* and *7''*, locate the points *f*, *g*, and *h*, which are equal to the length of *i-f*, *j-g*, and *7'-h*, shown at the top of the sheet.

STEP 11. Using a *straight edge* at points *f*, *g*, *h*, carry the diagonal lines across the vertical lines which extend down from the **tee**. This establishes the points *2''*, *3''*, *4''*, *5''*, *6''*, as shown in Fig. 7-40.

Round Layout Development

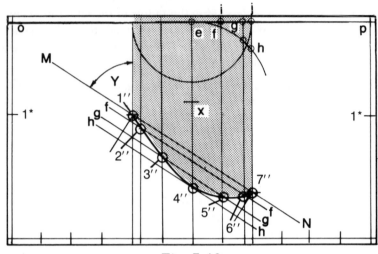

Fig. 7-40

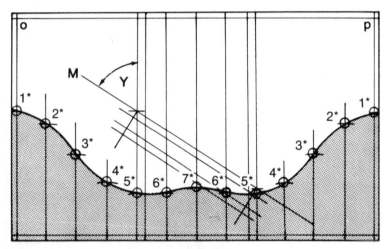

Fig. 7-41

STEP 12. Using a *square* off the bottom of the sheet, extend the points *1* to *7* of the base line upward to about the height of point *1**, as shown in Fig. 7-41.

STEP 13. Using either *dividers* or *trammel points*, transfer the lengths from the points *1″* through *7″* inclusive down to their respective places on the base line to establish the points of the *miter line*. These are indicated in Fig. 7-41 by the following: *1**, *2**, *3**, etc.

STEP 14. Connect points to form a continuous *miter line* and cut on this same line.

TEE NO. 6. A Tee or Pipe Intersecting Another Pipe of a
Different Diameter, on a Tangent, at an Angle

STEP 1. Let the rectangle *A, B, C, D*, shown in Fig. 7-42, represent the blocked-out metal required for the **tee**.

The length *A-B* equals the circumference, plus the seam allowance. The height *A-C* equals the longest point of the **tee**, plus 2″ or 3″ working area.

STEP 2. Scribe a 1/2″ line completely around the sheet. The lines on each end represent seam allowance, with the top and bottom being used for *layout* purposes.

STEP 3. Divide the **base line** into 12 equal spaces; number as shown in Fig. 7-42. At the center of the base line or point 7, use a *square* or metal *T-square* off the bottom of the sheet to erect a perpendicular line to the top of the sheet, locating point *x*.

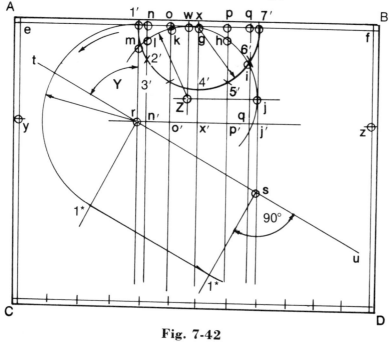

Fig. 7-42

STEP 4. With the *dividers* set to the radius of the **tee** and using *x* as the radius point, scribe a half-circle which locates points *1′*, *4′*, and *7′*. Then, in the usual manner, determine the points *2′*, *3′*, *5′*, and *6′*.

STEP 5. Again using the *square* off the *bottom* of the sheet, extend

Round Layout Development

the points *1'* to *7'* vertically to the top of the sheet to establish the points *n, o, p, q* shown on the line *e-f* in Fig. 7-42. Also extend these same lines downward almost to the base line.

STEP 6. Measure down from point *1'* using the desired length, to locate point *r*. From this point, place a diagonal line, *t-u*, using the required degree of angle, as indicated at *Y*. Line *t-u* represents the outside of the intersected *pipe* and also locates points *s*, as shown in Fig. 7-42.

Because the drawings shown here are reduced down to such a small size, it is most difficult to show any decided difference between the **tee** and the *pipe* it intersects. This will become more apparent as the *layout* progresses. However, in actual *shop* practice, the *fittings* will be large enough to require the developing of each point on the *layout*.

STEP 7. Using the *radius* of the larger intersected *pipe* and point *7'* as the radius point, strike an arc across line *e-f*, locating point *w*; then bisect line *7'-s* which locates point *j*, as shown in Fig. 7-42.

STEP 8. Again using the **same** radius and with points *w* and *j* as radius points, scribe intersecting arcs to establish point *Z*, Fig. 7-42. With the *dividers* still set the same, and using point *Z* as the radius point, strike an arc completely through the **tee** *layout*. This will establish points *m, l, k, g, h,* and *i* on the curve, as shown. This part represents the **tee** intersecting the *pipe* and is a view rotated 90° from the portion below the diagonal line *t-u*.

STEP 9. Place a *square* on the diagonal line *t-u* at points *r* and *s* and, at 90°, draw lines out slightly longer than the length *1'* to *r*, as shown in Fig. 7-42.

STEP 10. Again from point *r*, carry a horizontal line across the vertical lines of the intersecting **tee**, thereby locating the points *n', o', p', q',* and *j'*.

STEP 11. The *layout* process is now ready for development of the *miter line* points below the diagonal line *t-u*.

From point *r*, establish point *1**, which is actually point *1'* carried around. Also locate point *1** down from point *s*, as shown in Figs. 7-42 and 7-43.

Neither the arc from point *1'* to *1** nor the line from *1** to *1** is required when making the *layout*. Their only purpose here is to show the origin of these points.

The various lines must now be transferred to their respective

166

places on lines *r-1** and *s-1**, Fig. 7-43. This process of transferring the lines from above the diagonal line *t-u* to below it can be handled in 3 different ways: from top down, bottom up, or 1/2 down and the other 1/2 up.

It should be understood, before proceeding with the final steps of the *layout* that line *r-j'* and point *1** are only required when the various points are to be taken off this line. Then the points *n, o, p, q,* which were developed in **Step 5,** are not required, since they are only necessary when points are taken off top line *e-f,* Fig. 7-43.

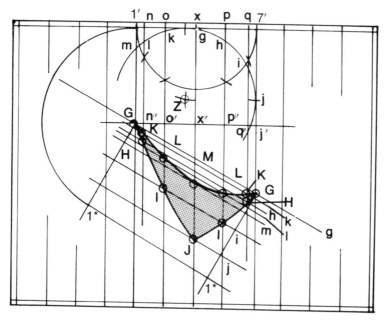

Fig. 7-43

To explain more fully: the lines formed between the points *l'-m, n-l,* and *k-o* are taken from the line *e-f* and, if transferred in the same manner, would be measured downward off the diagonal line *t-u.* If they were to be taken off the bottom line *r-j',* the lines would then be taken from between the points *m-r, l-n',* and *k-o'.* If transferred in this manner, they would be measured up from the line *1*-1*,* which is the *method* used here in Fig. 7-43. The same situation occurs here on the other half of the **tee.** These lines formed between the points *x-g, p-h, q-i,* and *7'-j* are transferred down in the same manner and therefore, are taken downward, off the diagonal line *t-u,* when taken off the line *e-f,* or upward, off the line *1*-1*,* when taken off the line *r-j'.*

Round Layout Development

STEP 12. For this particular problem, in which the **tee** is so small in size, the necessary lines for developing the **miter line** will be taken off the line *r-j'*. Each set of points to be measured upward from line *l*-1**, Fig. 7-43. Placing the points up both lines makes it easy to scribe the intersecting lines, which are used in developing the points for laying out the **miter line.**

STEP 13. With the intersecting lines established, it is necessary to determine the points which will, when laid out in successive order, make up the contour of the **miter line**, Fig. 7-44.

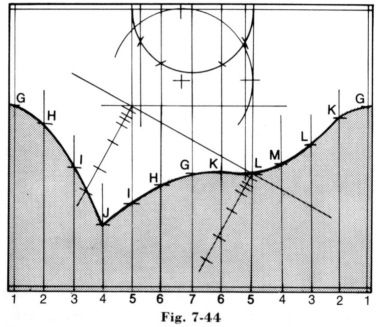

Fig. 7-44

To correctly locate each point, it is necessary to carry each point over horizontally to the particular line on which it belongs.

STEP 14. When placing the lettered points on the vertical lines, the *cutter* must be certain he carries each one to its proper place.

The development procedure used for this particular *fitting* is rather deceiving because the intersected *pipe* is placed a quarter of a turn around from its actual position.

However, by placing the lowest point at *4*, and knowing that this point (which is indicated here as *j*) is on this side of the **tee,** it is a simple matter to complete the *layout*, as each letter will follow in successive order.

The completed curve of the **miter line**, as it should look with

168

each letter in its respective place, is shown in Fig. 7-44.

STEP 15. Use every precaution in both the drawing of the **faired** line for the *miter* and in cutting out on this same line to complete the *fitting*.

TEE NO. 7. A Tee or Pipe Intersecting a
                Symmetrical Taper, at an Angle.

Although this particular **tee** is shown here being developed in the usual manner, it may not always be convenient to do so.

Since this *layout* involves considerably more drawings in order to develop the **miter line**, a larger sheet of metal may be necessary to do the actual development. Should the *fitting* in question be very large, it might be necessary to do the *pattern layout* on a separate piece of metal. However, this does not change the actual *layout* procedure used here, except for not being able to *lay out* the **tee** directly on the blocked-out material.

STEP 1. Let the rectangle *A, B, C, D*, shown in Fig. 7-45, represent the blocked-out metal required for the **tee**.

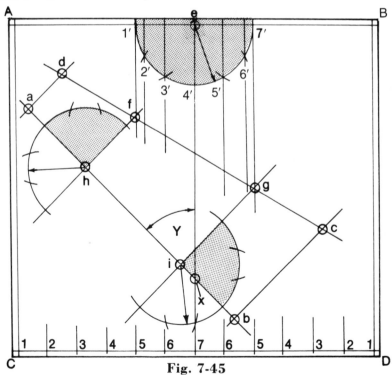

Fig. 7-45

Round Layout Development

The length *A-B* equals the circumference, plus the seam allowance.

For this particular **tee**, it is necessary to *lay out* one-half of the *taper* in order to develop the *miter line* of the **tee**. Therefore, the height of the blocked-out material has to be considerably longer than that actually required for the **tee** itself. It is usually satisfactory to use the full width of the sheet.

STEP 2. Scribe a 1/2″ line completely around the sheet shown in Fig. 7-45. The lines on each end represent seam allowance, with the top and bottom being used for *layout* purposes.

STEP 3. Divide the **base line** into 12 equal spaces; number as shown in Fig. 7-45. At the center of the base line or point 7, use a *square* or metal *T-square* off the bottom of the sheet to erect a perpendicular line to the top of the sheet, locating point *e*.

STEP 4. At a convenient distance above the base line, locate the point *x* on the center line.

Crossing the center line, *e-7*, diagonally at the required *angle*, determined at *y*, draw in the center line *a-b*, which represents the center line of the intersected *taper*.

STEP 5. With the *dividers* set to the radius of the **tee** and using *e* as the radius point, scribe a half-circle which locates points *1′*, *4′*, and *7′*. Then, in the usual manner, determine the points *2′*, *3′*, *5′*, and *6′*, Fig. 7-45.

STEP 6. Using a *square* from the bottom of the sheet, extend the points *1* through *7* inclusive, about half-way down the sheet. Knowing the length desired for the **tee**, measure down from points *1* and *7* to establish points *g* and *f*.

STEP 7. Placing a long *straightedge* at points *f* and *g*, scribe a line representing the outside of the *taper*, Fig. 7-45. From point *f*, measure up on this same line (or straight if necessary) to locate point *d* at the top of the *taper*.

STEP 8. Using a *square* off the center line *a-b* and touching point *d*, scribe a line which is the top of the *taper*. The distance between *a* and *d* should be equal to one-half the diameter of the *taper* opening. If this measurement is incorrect, the center line should be moved either way, as needed.

Refer to **Step 4**, where no definite location of this line was given at that time.

STEP 9. Knowing the exact length of the *taper*, the bottom of the *taper* is now established by measuring down from point *a* on the center line, thereby locating the correct point *b*.

Square off from this point *b* and, if the layout is thus far correct, point *c* will be 1/2 of the diameter of the bottom of the *taper* when measured from point *b*, Fig. 7-45.

STEP 10. Using a *square* off the center line *a-b* and placed against points *f* and *g*, the points *h*, *i* are established, as shown.

STEP 11. With the *dividers* set equal to the radius of the **tee,** and using points *h* and *i* as centers, scribe full half-circles. With the same setting on the *dividers*, follow the usual *method* of dividing the half-circles into six parts.

STEP 12. The half-circles are used as a means of convenience enabling the use of a *straightedge* to carry the lines over more accurately than would be possible with a *square* off the center line.

Using points *h* and *i* as centers, scribe arcs from points *f* and *g* at a random length. The divider points of the half-circles are now carried straight across to intersect these arcs in Fig. 7-46.

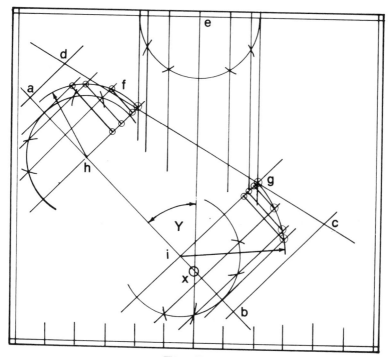

Fig. 7-46

Round Layout Development

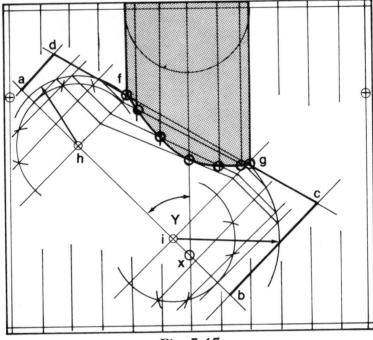

Fig. 7-47

These points are now squared down to lines *h-f* and *i-g*. By the connecting lines carried across the vertical lines of the **tee**, the *miter line* points are established in Fig. 7-47.

STEP 13. With the points of the *miter line* established (a connecting line through these points is not necessary), the usual manner of *laying out* the full *pattern* of the **tee** is now used to complete the *fitting* as shown in Fig. 7-48.

For this particular *fitting*, it is usually easier to develop the *miter line* by taking the points off the top rather than from the bottom, as has been done in each of the other **tee** *layouts.* As usual, the seam will be in the throat, shown as *l-f* in Fig. 7-48.

The shaded area in Fig. 7-47 is there to show the actual **tee** as it appears when set in place.

The shaded area shown in Fig. 7-48 represents material that is to be cut away. As usual, it is not necessary to allow any material below the curve of the *miter.* The faired line must be carefully developed as it will also be the cutting line.

Note that *g*, at the center of the developed *miter line*, is the *g* taken from the elevation of the **tee**. Thus, the various points are taken in the same order, up to the seam line or *f*, which is shown as *F* at each end of the *miter line*, Fig. 7-48.

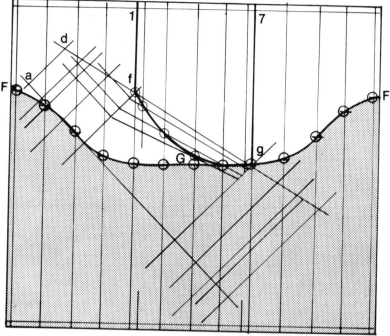

Fig. 7-48

INSTALLING THE TEE

The usual *method* of attaching the **tee** to the *pipe* is to notch or dovetail the edge of the **tee**. Then, with one tab in and one tab out all around the **tee**, place it into the opening in the *pipe* and bend the tabs tightly against the sides of the *pipe*.

It is easier to install the **tee** if the hole is prepared as shown in Fig. 7-49. Leave about 2 or 3 inches of stock on the top of the hole and cut the same amount off the edge of the **tee**, as indicated at *b* in Fig. 7-50.

As this edge around the **tee** is usually soldered, it is possible to do this on the inside, at the throat section, with the balance of the soldering done on the outside, as usual. To solder the throat from the outside is usually a difficult task, as the **tee** is often at a sharp angle, making it almost impossible to reach with a soldering iron. The shaded area of *k* is cut out. For edge *a* around the **tee**, use either the *thickedge, elbow machine* or *crimper* to make flange conform to the pipe and pop rivet in place.

Round Layout Development

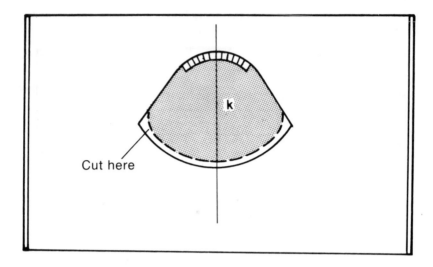

Cut here

k

Fig. 7-49

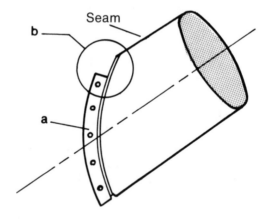

Seam

b

a

Fig. 7-50

WYE BRANCH FITTING

This type of *fitting*, often referred to as a *pair of pants*, is not actually an *air conditioning fitting*. However, it is often used, along with the various **tees**, in high-pressure *air conditioning systems* using *round pipe*.

Although not shown here, a similar type *fitting* having three branches is also made in the same manner.

Shown in Fig. 7-51 is an elevation of the section to be laid out,

174

marked *D*, which is half of the *fitting*. A lap across the center from *X* to *14* is only required on half the *fitting*.

STEP 1. Having laid out the *elevation D*, scribe the half-circle *C* and two quarter-circles *A* and *B*, Fig. 7-51. Divide into equal parts, as shown, number each point and connect with solid or dotted lines.

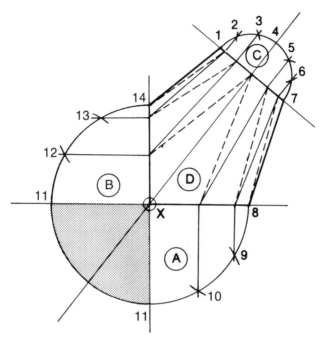

Fig. 7-51

STEP 2. Determine the true length of these lines, as shown in Figs. 7-51 and 7-52. This *method* of *triangulating* the lines is the same as used in the development of **round roof jack** (to a pitch); see **Index**.

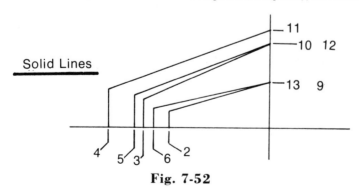

Fig. 7-52

Round Layout Development

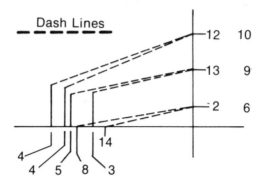

Fig. 7-53

STEP 3. The usual manner of *layout* is applied, beginning with line *7-8* which is a true length. The seam will be on line *1-14* which is the throat and is also a true length line.

STEP 4. An allowance must be made on the smaller openings for *collars*, since the air will most likely be moving in that direction.

For the larger opening, make a heavy gauge *collar* and have the allowance on the *collar* to fit inside. These *collars* should be spot-welded on when possible.

Care must be used in a *layout* of this type, as the *pattern* will have a tendency to grow, thereby making the opening too large. The circumference of each opening should be very carefully checked while the *pattern* is in the flat, Fig. 7-54.

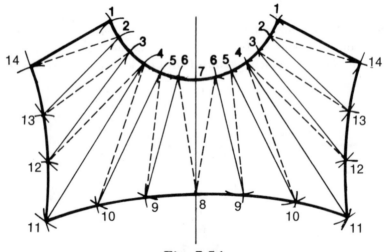

Fig. 7-54

A *layout* like this requires considerable time and most *cutters*, for this reason, will use the **roll-out** *method*, also shown here. This *method* is considerably faster and, if used carefully, is quite accurate.

Shown in Fig. 7-55 is the *layout* of the *fitting* as it appears before making the breaks that are required for the rolling process. This *method* of *layout*, the development of which is described in the next section, should appear the same as that shown in Fig. 7-54, which was laid out in the regular manner. The various points, as shown on the *pattern* in Fig. 7-55 as follows: *11* and *f*; *8* and *e*; *1* and *c*; *7* and *d*; and *14* and *b*.

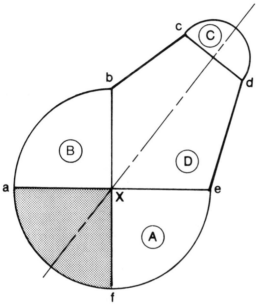

Fig. 7-55

WYE BRANCH FITTING
Roll-out Method

This *method* of *fitting development* is used quite often and, although it could hardly be considered **the** correct way, it can be applied with amazing results.

The **roll-out** process is quite simple: however, in order to develop a good usable *pattern*, it is necessary to follow through with each individual step most carefully.

Theoretically, the metal *pattern* or *template* is assumed to repre-

Round Layout Development

sent the inside as solid. It might also be expressed as being a finished *fitting*, the metal *pattern* to be developed by wrapping the metal around this in order to obtain the desired shape.

This *method* is, therefore, applicable to any similar type of *fitting* such as a *square to round, three-branch fitting, roof jack*, etc.

STEP 1. Lay out the actual *elevation* of the *fitting* in the usual manner, Fig. 7-55. Also draw both the half and quarter-circles which are indicated by the shaded area. The *layout* should be made on either 22 or 24 gauge iron, making it easier to hold in shape throughout developing process.

STEP 2. Use extreme care in cutting out the *pattern*, as this edge is what actually makes the *layout*. The half and quarter-circles *A, B, C*, Fig. 7-55, are turned up 90° at *c-d, b-x*, and *x-e*. It is important that these breaks remain perfectly square and, if necessary, the turned up ends should have added support in the form of *gussets* or some similar type of brace. It is usually only on the very large *fittings* that some form of stiffener is required to insure holding the ends square.

STEP 3. The formed section is now ready to roll out. Using either *asbestos* paper, which shows the imprint when rolled out on it, or by using regular paper with a pencil, the outline can be developed on the paper, forming a half of the *pattern*. This, in turn, can be cut in full *pattern* by merely folding the *half-pattern*; then by tracing around this *pattern* onto the metal, the *fitting layout* is almost complete, requiring only edge for a seam. Also add the allowance on one side for the necessary lap in order to complete the *fitting*.

When the metal *pattern* is complete, a second *fitting* is required. The lap which is to tie the two *fittings* together is only required on one of the *fittings*.

It is quite obvious that, in rolling out this *fitting*, it will grow to some extent. Therefore, it is a good practice for the *cutter* to check the *stretchouts* upon completion of the *layout*.

Refer back to Fig. 7-54. Notice that, in this regular *layout* of the **wye** *fitting*, the lower part of the *layout*, or points *11 to 11*, is one-half of the *stretchout* of the larger opening.

ROLLING HEAVY METAL

When **rolling heavy metal** for *pipe* or *fittings*, it is necessary that allowances be made to compensate for shrinkage of the material.

178

As shown in Fig. 7-56, three *different diameters* are obtained when rolling heavy metal: the **inside** diameter, *B*; the **outside** diameter, *C*; and the center or **mean** diameter, as it is commonly called, *A*.

When rolling heavy metal, we are required to hold to either an **inside** or **outside** diameter. To do this, the metal thickness must be determined and allowance made as required. To obtain either **inside** or **outside** *diameters*, the actual *stretchout* is made on the **mean** diameter.

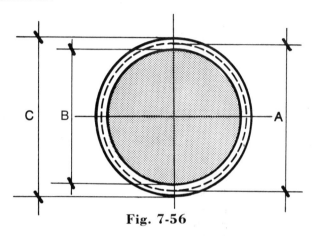

Fig. 7-56

To aid in the explanation of this: assume the metal thickness, shown in Fig. 7-56, to be 1/8'' thick or 1/16'' on both sides of the **mean** diameter, *A*. Therefore, if the *stretchout* for a 12'' diameter was laid out on the metal, with no allowance:

    A. **mean** diameter would be 12''

    B. **inside** diameter would be only 11-7/8''

    C. **outside** diameter would increase to 12-1/8''

The actual *stretchout* for a twelve-inch diameter would be 37-11/16''. To obtain an **inside** diameter of 12'', using 1/8'' metal, it would be necessary to use the *stretchout* for a diameter of 12-1/8'', which is 38-1/16''.

To obtain an **outside** diameter of 12'', it is necessary to use a *stretchout* for a diameter of 11-7/8'' which, using 1/8'' metal, is 37-5/16''.

To sum up briefly: to obtain the **inside** diameter, **add** one metal thickness; to obtain the **outside** diameter, **subtract** one metal thickness.

The so-called **mean** diameter is actually what is regularly used for the metals of lighter gauge. However, the metal's growth or

Round Layout Development

shrinkage is either too small to be considered or is of no importance, since the *pipe* or *fittings* can be crimped to fit correctly.

For heavy *pipe* that is to be riveted together, it is necessary to have a **large** and a **small** end. The small end is seven times metal thickness, deducted from the circumference of the required diameter.

The *shearing* presents somewhat of a problem when a small end is required on a pipe. However, the pieces can be cut out at an angle by cutting out the first piece and using it as a *pattern*, Fig. 7-57. Then, scribe each edge of the sheet and sight these marks at the blade edge of the *shear*. Always be certain that the required lap for the longitudinal seam has been included in the *pattern*.

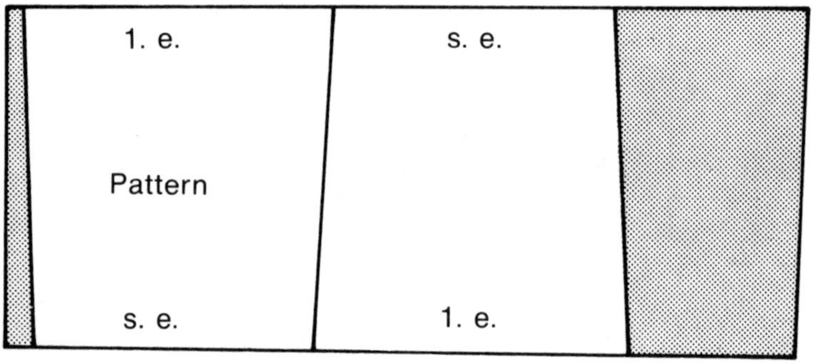

Fig. 7-57

Marking the large and small ends on each piece, as it is being sheared, is a time-saver, Fig. 7-57, since the difference between the ends is usually hard to spot at a glance, particularly when a large amount of pipe is sheared at one time.

NOTE: When shearing heavy metal, be certain of the shear's capacity. At any rate, do not attempt to shear heavy material too close to the point where the blade starts to drop.

ROUND LAYOUT WITH HEAVY METAL

The specifications for making up **boiler breeching, stacks,** etc., call for heavy gauge metal such as 10, 12, or 14 gauge *black iron.* This requires welded construction throughout, except for the field joints which are usually made with *companion flanges* *although, in some cases, they too, are welded on the job.*

In almost all cases, a sketch or plan of the work required is checked out at the *jobsite* and sent in to the *shop*. This may be a scaled *drawing* or nothing more than a rough *sketch:* this is not important, as long as the necessary measurements are shown, such as overall dimensions, center-to-center, etc. It does not have to show the lengths of the *fittings*, as these can be determined in the *shop* (unless there are certain *fittings* which must be kept within a specified length).

It is the usual practice in most *shops*, whenever there is a complete section of *round fittings* and *pipe*, such as referred to here, for one *cutter* to work up the entire *job*, making up the *shear list* and *patterns*, with the development and responsibility being in his hands.

The *reader's* attention is called to the fact that what is to be described here is actual *shop* procedure as it should be applied. All the calculations and *methods* of procedure are those used ·by the *author* and can be applied to other large *layouts in the same manner.*

As a means of better understanding how to handle the development of a *layout* such as this, assume the *sketch* shown in Fig. 7-58 to be the *layout* described on the following pages.

Before starting to prepare the *shear list*, it is important to check the actual **length** and **width** of the material to be used. The metal will undoubtedly be the regular 48″ x 120′′ sheet which is the most commonly used size. These sheets almost always run both slightly wider and longer. Although the width will only be about 1/8″ wider, the length usually runs up to about 1/2″ longer. It is also important to note if the sheets are *square* enough to be used as they come, without having to make a trim cut to *square-off* the sheet.

In either case, a definite width and length should be established that is usable as a full sheet and this exact size kept in mind when preparing the *shear list*.

So that the development of Fig. 7-58 can be described here, it will be assumed that the metal to be used here is 48-1/8″ x 120-1/2″, is 10 gauge *black iron*, and that the sheets are *square* enough to be used at this size without requiring a trim cut.

STEP 1. Check over the entire *layout* to see if all the necessary dimensions are shown: also make a check on these same dimensions by adding the center-to-center lengths against the overall length. These should measure out the same. Any discrepancy that may appear should be traced to the source of the error and corrected

Round Layout Development

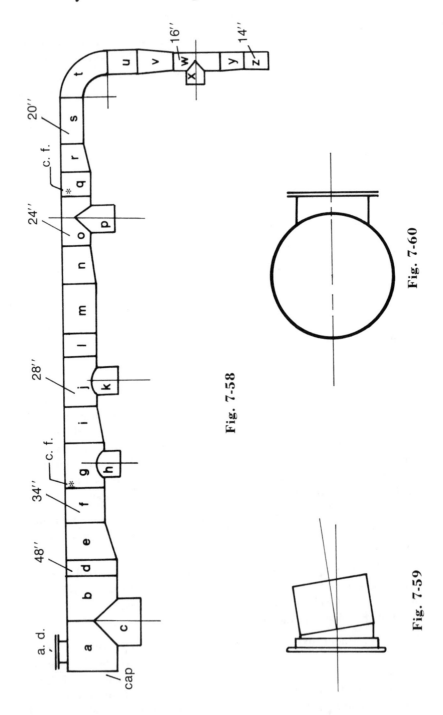

Fig. 7-58

Fig. 7-59

Fig. 7-60

before attempting to prepare a *shear list*. It is always assumed that the center-to-center dimensions are critical and must be maintained, since these are established openings whose location cannot be changed. However, it is unwise to attempt to guess which figures are to be used and the *cutter* should not assume this responsibility.

STEP 2. When all calculations check out correctly, the next step is to determine the gauge of metal required and whether the diameters shown are to be **inside diameters** or **outside diameters.** This information is generally given with each *job*. For the hypothetical *layout* shown in Fig. 7-58, assume the metal to be 10 gauge *black iron*, with the diameters to be **inside,** or ID as it is abbreviated.

STEP 3. Refer to the *plan* in Fig. 7-58 where the following diameters are shown: 48'', 34'', 28'', 24'', 20'', 16'', 14''. All are to be ID and the correct stretchouts are derived in the following manner: Refer to the **Circumference Table** for the *stretchouts* of each required size. This will give the **mean diameter** *stretchouts*, which are explained in this *Chapter* in the section, **Rolling Heavy Metal.**

STEP 4. Refer to **Math Problem No. 20** and note the *factor* .442 opposite 10 gauge. Since the *inside diameter*, ID, is required here, this *factor* is **added** to each of the *circumference stretchouts* shown with the following results.

DIA.	CIRCUM.	FACTOR		S. O.	WORKABLE FRACTION*	
48''	150.796	+ .442	=	151.238	(15/64)	151-1/4''
34''	106.814	+ .442	=	107.256	(1/4)	107-1/4''
28''	87.965	+ .442	=	88.407	(13/32)	88-7/16''
24''	75.398	+ .442	=	75.840	(27/32)	75-7/8''
20''	62.832	+ .442	=	63.274	(9/32)	63-5/16''
16''	50.265	+ .442	=	50.707	(23/32)	50-3/4''
14''	43.982	+ .442	=	44.424	(27/64)	44-7/16''

* A *workable fraction* refers to one that can be read on the *rule* in sixteenths of an inch, since this is the smallest division of the *fraction* used in the *sheet metal trade* (with a few exceptions).

NOTE: When an **inside diameter** *(ID) circumference stretchout is required, select the closest workable fraction* **above** *the one shown in the parentheses. When an* **outside diameter** *(OD) circumference stretchout is required, select the closest workable fraction* **below** *the calculated fraction shown in the parentheses. The above figures are determined by means of a Circumference Chart and a Decimal Equivalent Chart, both of which are in the Chart Section of this book.*

Round Layout Development

Notice that in each case where a workable *fraction* is selected, it is only a difference of 1/32" or 1/64" and, used in this manner, it is not necessarily critical. However, if the procedure were reversed, it would make a difference in the workable *fraction* of about 1/16" in practically every final calculation shown under Workable Fraction. The point here is to avoid any possible difficulty by applying the *method* just described. It is a simple matter to reason that, if an OD *stretchout* is required, then the *diameter* can be slightly under the *mean diameter*, but **not** over. The same logic applies if an ID *stretchout* is required, the *diameter* can be slightly over the *mean diameter*, but **not** under.

Working with heavy metal requires careful handling at all times and it is important that, once these figures are worked out, no changes should be made. The same procedure is used in all cases, with the **gauge of metal** determining which **factor** is to be used.

Although every attempt is made to calculate these figures as accurately as possible, there is no assurance that this accuracy can be maintained throughout the process of *shearing* and *fabrication*. However, one can usually take care of any discrepancies that may occur when the *fitting* is being *welded* together.

Because it is important to be able to calculate the **outside diameter** as well as the **inside diameter** *circumferences*, they are also shown here, although they have **no** part in the *layout* being developed in Fig. 7-58. The purpose here is to show both *methods*, with the results of each. This will give the *reader* a better understanding of the differences between the ID and OD *circumference stretchouts*, which are the result of **rolling heavy metal.**

Refer again to **Math Problem No. 20** and note that the same *factor*, .442 for 10 gauge, is also used to determine the OD shown here. This *factor* is **subtracted** from each *circumference*, which is opposite procedure from that applied in **Step 4.** The results are as follows:

DIA.	CIRCUM.		FACTOR		S. O.	WORKABLE FRACTION	
48"	150.796	−	.442	=	150.350	(11/32)	150-5/16"
34"	106.814	−	.442	=	106.372	(3/8)	106-3/8"
28"	87.965	−	.442	=	87.523	(33/64)	87-1/2"
24"	75.398	−	.442	=	74.956	(61/64)	74-15/16
20"	62.832	−	.442	=	62.390	(25/64)	62-3/8"
16"	50.265	−	.442	=	49.823	(53/64)	49-13/16"
14"	43.982	−	.442	=	43.540	(35/64)	43-1/2"

184

Layout with Heavy Metal

This *method* of determining the *circumference stretchout* for both the ID and OD can be applied in the same manner when using other gauges of metal. It is only necessary to use a *decimal factor* that applies to the gauge of metal being used. These factors can be found in **Math Problem No. 20.**

STEP 5. Having established the correct *stretchouts*, a list of all *pipe* and *fittings* is taken off the *plan* shown in Fig. 7-58. The following is a breakdown of Fig. 7-58 which includes all the *pipe* and *fittings*. Note that each piece is identified by a letter rather than a given number, which is the author's usual *method*. The only purpose is to avoid possible confusion with the numbers that are shown for the given *dimensions*.

Breakdown for Fig. 7-58

a	= full jt.	48'' pipe	*n	= taper	28'' to 24''
b	= full jt.	48'' pipe	o	= full jt.	24'' pipe
*c	= tee	48''	*p	= tee	24''
d	= 18'' stub	48'' pipe	q	= 20'' stub	24'' pipe
*e	= taper	48'' to 34''	*r	= taper	24'' to 20''
f	= 36'' stub	34'' pipe	s	= full jt.	20'' pipe
g	= full jt.	34'' pipe	*t	= elbow, 5 pc.	20'', 14'' rad.
*h	= tee	16''	u	= 42'' stub	20'' pipe
*i	= taper	34'' to 28''	*v	= taper	20'' to 16''
j	= full jt.	28'' pipe	w	= full jt.	16'' pipe
*k	= tee	28''	*x	= tee	16''
l	= 36'' stub	28'' pipe	*y	= taper	16'' to 14''
m	= full it.	28'' pipe	z	= 36'' stub	14'' pipe

*Requires *light gauge galvanized iron* **pattern**

The foregoing list includes all but the 48'' dia. *cap* for *pipe a* and the *access door.*

It is assumed here that the *layout* shown in Fig. 7-58 is to be made up into three *sections.* This will require two *field joints, as indicated* between *fittings f* and *g,* and *o* and *q,* which can be made with two sets of *companion flanges.* These are usually ordered from a *shop* that specializes in making this type of *flange.* To be certain of obtaining the exact size required, it is a good practice, when ordering, to state the gauge of metal used, whether ID or OD diameter, and not just ask for the diameter size only. These rings or *flanges* will have an excess of 6 or 8 inches of *angle iron* which must be cut off. It is best to use a long steel tape and measure the girth of the *pipe* or *fitting* that the *flange* will have to fit. After *cutting* the *flanges*

Round Layout Development

and *welding* the joints, the bolt holes can be laid out. The best means of doing this is to make a pattern on light gauge galvanized iron, with the *hole layout* being carefully developed. It is advisable to transfer these hole centers from the *pattern* to only one *flange*. Then, drill through both the *flanges* at one time by holding them together, back-to-back, with several pairs of vise-grips. Always put through a small pilot hole, then follow up with the required size of drill which will be determined by the size of bolts being used. Usually, 3/8'' bolts use a 7/16'' hole.

STEP 6. Attention is called to the fact that the various *fittings* and *pipe* are given an exact length in the **breakdown** taken from the *sketch* in Fig. 7-58; however, these will not necessarily be the lengths used on the *shear list* which is now prepared.

Theoretically, since the *pipe* and *fittings* have raw edges and are merely butting together, it might be assumed that there is very little, if any, growth along the line. However, any irregularities in the *cutting out* of the *fittings*, or some small burrs on the edges, are bound to cause some growth along the *line*. Growth can also be caused by carelessness during the fabrication, such as failure to draw the *pipe* and *fittings* close enough before *welding* in place. Since this growth is variable and unpredictable, it cannot be set at a definite figure. It is usually safe to figure about 1/8'' per joint and, if any adjustment is necessary, it can be made at the *companion flanges* or the last joint of the assembled *line*.

As for the critical *center-to-center* location, which occurs between the **tees**, such as *a* and *k* in Fig. 7-58, it is advisable **not** to attempt any hole cutting for these *tees* until the *line* is otherwise satisfactorily completed. This makes it possible to establish exact locations of the various *takeoffs*. With the main trunk line put together, it is easy to measure off for the *tees* and mark out the holes, which can be cut out with a *cutting torch*.

It is best to make up a *cutting sheet* and show each *fitting* in a rough outline. Although it is not necessary to *scale* them out, they should be somewhat proportioned in the general direction of the diameter change. This makes it easier to visualize, at a glance, just what type *fitting* is required.

In most cases the length to which each *fitting* is to be made will be indicated on the *plan*. However, in some places, the length will be omitted. It is usually assumed that, if this length is not shown, the length of the *fitting* in question is not critical and it is left to the *cutter's* discretion to develop a *fitting* that will not only look good

Layout with Heavy Metal

but be properly proportioned.

The various *fittings* shown in Fig. 7-58 require *patterns* made of light gauge galvanized *iron* and development of these *fittings* is shown and explained in other parts of this *Chapter*. Therefore, there is no purpose in repeating them here. The material for these *fittings* will be made from the sheet *drop-offs* resulting from the *pipe block-outs*. The first step in cutting the material is to block out the pipe and prepare a *shear list* for the *shearman*. While this is being blocked out, the *cutter* can develop the *patterns* for the *fittings*.

The *reader* is reminded that, in making the galvanized *patterns*, the *stretchouts* must be taken from the calculations made for this *layout*. Since the *fittings* will be made up out of the *drop-offs* of material, it will be necessary, in most cases, to make the *fittings* in half-sections. This is usually a good practice, regardless of this fact, since trying to cut such *fittings* as *tapers* and *tees* in one piece is not very practical. When laid out in the flat, these *fittings* make such irregular shapes that they waste a considerable amount of metal. The cost of *welding* the extra *seam* does not come near the cost of material involved or the extra *layout* time. Furthermore, these *fittings*, when cut out of one piece of 10 gauge black iron, are too heavy to handle. It is not too bad with the very small *tees* or *tapers*, as they are about equal to a half of one of the larger *fittings*.

Due to the fact that the *stretchouts* are already established, and no other allowance is required except the 1/8'' referred to regarding growth, preparing the *shear list* is a simple operation. As all of the *stretchouts*, except for the 48'' diameter pipe, are within the sheet size, there is no need to show a *shear list* for the *pipe*. However, an example of how to consider the 48'' diameter *pipe* blockout is shown here.

On checking the *circumference* for the 48'' ID, as calculated in **Step** 5, the *stretchout* is 151-1/4''. Use the full length of the sheet, 120-1/2'', which, subtracted from the *stretchout* of 151-1/4'', will be 30-3/4'' short of the required length. Although it will require two *welded* seams to make up the *pipe*, 1/8'' will usually be sufficient allowance for growth when added in this manner. Therefore, the blockout sizes for a joint of 48'' diameter *pipe*, as it is being used here, should appear on the *shear list* as:

    1 pc. — 120-1/2'' x 48-1/8'' (full sheet)
    1 pc. — 30-5/8'' x 48-1/8''

It should be noted here that the 48'' *tee* should be made up exactly in half, both for convenience in cutting out and because it

Round Layout Development

is easier to roll and fabricate, and makes a better finished appearance when completed.

There are occasions when a long run of *pipe* and *fittings* is to be offset out from the *companion flanges*. As shown in Fig. 7-59, a short *mitered* piece should be developed in order to swing the *line*, as required, since there is no means of making any variation in the *line* while keeping the *companion flange* intact in the squared-off position.

If the amount of offset required is very slight, which is the case when a long line is concerned, then the added mitered pieces can be made from 0° at the throat to the amount necessary to make the offset in the line. It will not be necessary to miter the adjoining piece of *pipe*, since the slight variation can be absorbed when *welding* up the *joint* . The best way to determine how much the change must be is to make a *scaled sketch* of the *line* and work from this.

Fig. 7-60 is an enlarged view of the *access door* assembly, which consists of four angles being made and a flat block-out piece for the *door* being bolted in position. The bolts, usually 3/8″ dia., are welded in position to the underside of the frame and, in some instances, wing nuts are used to keep the door in place. Note in Fig. 7-60 that two sides of the *frame* must be fitted to the contour of the pipe, with the opposite sides being straight. The extended *frame* is usually kept about 2″ from the *pipe* itself.

chapter VIII
DUCT LAYOUT AND
RELATED FITTINGS

INTRODUCTION

Shearing, notching and **fabricating duct**, such as used in *air conditioning systems*, is more or less a routine process in the *sheet metal shops* of today. The *cutter* is seldom required to make up any *duct*, particularly in the larger *shops* where *duct* is handled as a separate product and is kept apart from the *cutting* department. However, the *cutter* should be able to make up *duct* when the occasion arises. While there is nothing difficult about it, it does require a certain *know-how* in regards to the necessary allowances to be made for the *machines*, how to prepare the *shear list*, etc.

With the *power equipment* that is available to *shear, notch, cut* and *hammer* the *duct* together, it can be understood why the cost per pound of *duct* is figured so low. In order to maintain this low manufacturing cost, the entire process involving the many phases required in turning out *duct*, from the *work sheet* to the *loading dock*, must be handled systematically.

Power equipment is to be found in even the smallest *shops*, since it would be impossible to even attempt to meet competition without it. A comparatively new *machine*, that measures, notches, and cuts off the exact length and size of *duct* from large *coils* of steel, will be found in the larger *shops* in the *industry*.

Preparing the *shear list* is the most important step in the entire process of making *duct*, since the **design**, **size** and **markings** are all determined, at this time, for each step that follows. It can therefore be understood that, unless the *shear list* is carefully prepared and the *shearman* uses good judgement and care in cutting the material, a tremendous loss can occur at this point, before the *duct* has even been notched.

It is for this reason that most of the *larger shops* have one man make up the required *shear lists* for all the *duct*. However, many of the shops have the *duct* and *fittings* on the same *work sheet* and the *cutter* does the entire sheet himself, including the *duct* (see Chapter on **Cutting with a System),** which is laid out at the *notching machine*.

Duct Layout and Fittings

It is a good practice, in this case, to make up the *duct shear list* first and, while waiting for this to be cut, the *cutter* can prepare his *shear list* for the *fittings*. He can then be notching the duct while waiting for his material for the *fittings*.

This also conforms with the common sense *method* of shearing the *duct* or largest pieces first, taking the smaller *fittings* from the drop-offs or scrap material.

Before making up the *shear list* for the *duct*, the following facts must be known:

      1. What gauge metal is required?
      2. What size sheets of metal are on hand?
      3. Is the *duct* to be panelled?
      4. Is **inside lining** required?

The actual design or *layout* of *duct* has its limitations. The length,

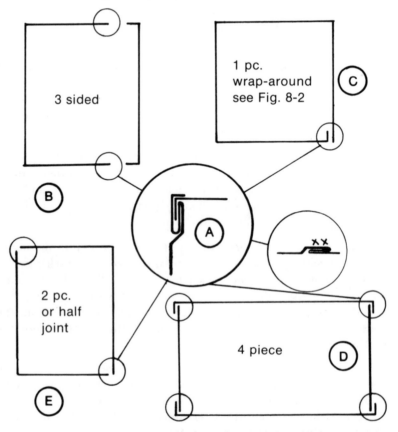

3 sided

1 pc.
wrap-around
see Fig. 8-2

2 pc.
or half
joint

4 piece

Fig. 8-1

of course, cannot exceed the length of the *brake*. The required size of *duct* will largely determine how the *layout* should be made and, when more than one way is available, the *cutter* must use his own discretion.

There are four ways in which the *duct* can be made, as shown at *B, C, D,* and *E* in Fig. 8-1. While it may be possible to form large size *duct* in one wrap-around section, or even in three-sided sections, this may be too large to handle. Under most conditions, the three-sided *layout* with the drop-in panel, as shown in *B*, will be the best for the larger size *duct*. With the *pittsburg* being on the three-sided section, *A* in Fig. 8-1, there is less handling at the *brake* and it is decidedly easier to fabricate. The *pittsburg* should always be placed on the **shortest** side, unless specified otherwise.

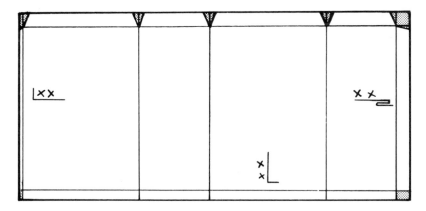

Fig. 8-2

Fig. 8-2 shows the *layout* for a wrap-around or one-piece joint of *duct*, as notched at the *bench*. The shaded areas indicate the cut-out sections. It is best, when making more than one joint of the same size, that a *pattern* be made without notching. The entire *layout* is prick marked, including the places that are to be notched out. By doing this, it is possible to mark through two or three of the joints at one time, then notching out as indicated by the prick marks.

Although a straight notch is all that is required on the flange end, when this notching is done by machine, both ends of the *duct* are notched with a *vee* cut. The same depth of notch, usually 7/8″, is used for both ends; whereas, when notched out by hand, it is customary to allow 1″ for the *lock* and 7/8″ for the *flange*.

Notice, in Fig. 8-2, that the *duct layout* has the *pittsburg* on the smaller side of the *duct* which is correct. Also note that the flange

191

Duct Layout and Fittings

is on the right-hand side in relation to the flange, which is on the bottom.

This is correct if the *duct* is to be formed on the *hand brake*. However, when the *power brake* is used, the *pittsburg* should be put on the other end, which merely requires that the marks be placed on the other side of the *layout*. The purpose for this being the fact that all small edges, such as 1/4″, are formed with the bulk of the sheet held out front when using the *power brake*. The small single edge is held into the *brake* against a back gauge or just held in place by sighting.

Inside insulation or **lining** of **duct** has become a widely accepted *method* of combining both **insulation** and **sound-proofing** into one process. When lining *duct* in this manner, it is best to cross-break or panel the *duct*, run through the *pittsburg* machine, turn up the 1/4″ edge and turn down the 7/8″ flange before applying the insulation to the inside. This keeps the *duct* flat, making it easier to apply the lining. After allowing a little time for the adhesive to dry, the final one, two or three breaks, whichever the case may be, are easily made to complete the *duct*.

When the *duct* is to be lined, the size of the *duct* as well as the job number, along with other necessary identifying marks, must be placed on the outside. The best place for these marks is on the 7/8″ flange, since this edge will remain visible until the *duct* is actually installed in place. After this, the marks serve no purpose.

When making *duct* with *S & D*, this can be notched on the notching machine by moving a pair of adjustment screws or pins, such as used on some *notchers*. The edge required is a half-inch, with the **drive** edge always being on the **short side** of the *duct*, unless it is otherwise specified.

Many substitute materials, such as *fiberglas*, *plastic*, etc., have been introduced as replacements for steel duct which is still being used to a certain degree. However, each of these materials has certain limitations, and it is the opinion of many in the trade that they seldom will satisfactorily replace standard galvanized iron duct.

Before the *pittsburg lock machine* became part of our trade, the *slip-lock* and *double-seam* were used as the means of making *duct*. However, this has long been gone and now, with both the *pittsburg* and *snap-lock* accepted by the *industry*, it is these two *locks* which are used exclusively.

Aside from requiring a different allowance for this type of **seam**, laying out **duct** which is to have the *snap-lock* is handled about the same as making the seam with a *pittsburg lock*. There is, however, one small allowance to be made as shown here in Fig. 8-3: 1/16″ is deducted from the side of the *duct* which has the single edge. The purpose of this is to take up the growth that builds up due to the bulk of the *locking* part of the *snap-lock*; or, as it is more commonly referred to, the female edge, *a*, the single edge being the male, *b* in Fig. 8-3.

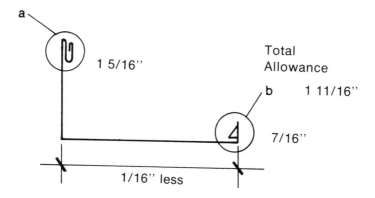

Fig. 8-3

Whether the *pittsburg lock* or the *snap-lock* is the better of the two is a matter of personal opinion, as they both make a neat looking *seam* and both are acceptable to the *industry*.

The *pittsburg* is the more versatile of the two. The outstanding difference between them is the fact that *pittsburg* seams must be closed by **using a hammer**, whereas a *snap-lock* seam is locked by merely inserting one piece into the other, then snapping into place **without using a hammer**. Although it can be done in just this manner, it is usually found, where long lengths and many joints have to be assembled, that a *mallet* or the side of a *hammer* is a most welcome aid in making that snapping-into-place process less tiresome.

The *locking* of the *snap-lock* is accomplished by means of the small raised *tabs*, indicated at *b* in Fig. 8-3, which are placed intermittently on the single edge, or *right-angle* bend, by the machine.

Fig. 8-3 shows an end view of both these edges which are formed in the machine. It is also shown that the difference in the gauge of metal requires more or less material. This also applies to the *pittsburg lock*. The various allowances required are to be found in the

Duct Layout and Fittings

Edge Allowance Chart shown in the Chapter, Cutting with a System.

The *snap-lock* closely resembles the *coffin lock* or *hidden seam*, which has been used in the *sheet metal trade* for as long as anyone can remember. The one variation between the two *seams* is the fact that the *snap-lock* is a self-closing or self-sustaining seam, whereas, the *coffin lock* has, by itself, no means of locking. The usual *method* of locking this seam is to tack-solder the seam along the inside.

OBSTRUCTION IN LINE OF DUCT

It is sometimes necessary to run a line of *duct* through an obstruction that cannot be avoided. Although this problem is seldom found in new *construction*, it is encountered quite often in remodeling work. This obstruction may be a pipe line or perhaps a supporting brace in one form or another. In either case, when the pipe cannot be removed or *duct* rerouted, the obstruction must be taken through the *duct*.

There are several *methods* of handling such a situation which, of course, differ with each type of obstruction and are based on certain *rules* regarding size of obstruction and the area loss which occurs.

In some cases it is possible to use a **split duct**; whereby the portion of the *duct* at the point of obstruction is divided to pass around each side of the obstruction. This is often referred to as an *easement* and can be made two ways. When the obstruction loss is under 20 percent, the *easement* can then be made with no change in the actual *duct* size. However, when the area loss is in excess of 20 percent, the area must be retained. This makes it necessary to make the offset section larger.

When the obstruction is close to the bottom of the *duct* line it can be passed, by using two top flat *transition fittings*. However, when using this *method* it is important to maintain a **rise** to the point of obstruction of about 1″ in 7″ but not less than 1″ in 4″.

The *fitting* leading away from this point should be in excess of 1″ in 7″. If the obstruction is not over 2″ and the duct in excess of 12″, the same size can be maintained (see dash lines, Fig. 8-4). However, when these figures are other than this, it is necessary to make the *transition fittings*, as shown.

One further point that should be given here is that the length of the shaded area shown in Fig. 8-4, *u-y*, should be kept not less than 3 times the diameter of the necessary enclosure, *k-k*. It is generally

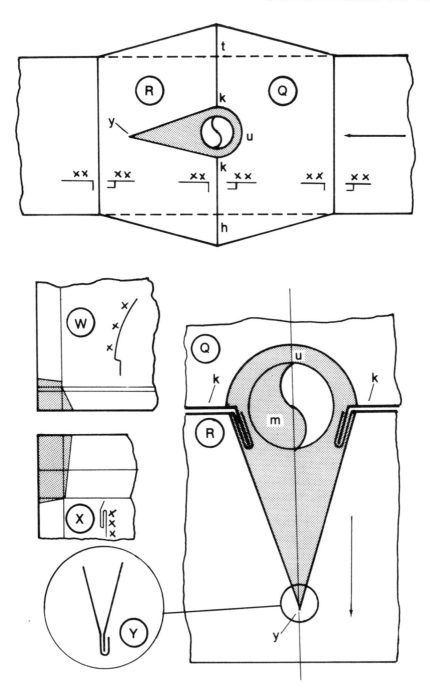

Fig. 8-4

Duct Layout and Fittings

accepted that an *easement* be used for round obstructions of 6″ diameter or larger; also for any square obstructions of 3″ or over. Where the obstructions are less than these figures, the *method* in Fig. 8-4 should be adopted: with *m* representing the pipe, sections *Q* and *R* must be laid out as shown.

The usual *lock* and *flange* can be used across *t-h*, being made as a three-sided government lock on each side of the pipe. Shown at *W* and *X* are corner *layouts* at the *S* lock used, with the shaded area showing the notching. The *S lock* is placed vertically at each side of the pipe, on the *vee* side of the connection. This is clearly shown at *Q* and *R* in the *blow-up* of the two sections cut for center area, the open area being shaded around the pipe *m*. The single edge of the connection, which is on section *Q*, is sharply offsetted to fit into the *S* forming the smooth surface on the inner side *k* when in place. The section as a whole is handled the same as throats of other *fittings*.

When it may be more convenient, make up the *vee* section in two pieces, a *standing seam* being used at the point as shown in detail *Y*.

VANES OR DUCT TURNS

Practically all *square elbows* require **vanes** or **duct turns,** as they are often called. However, they are often omitted in the return *air duct*, since air flow is not too critical in these lines.

As shown in Fig. 8-5, *vanes* can either be riveted directly to the cheeks of the *elbow, a,* or placed on a *track* and inserted into the *fitting*, as a unit, after fabricating the *elbow*. Marks *b* and *c* are for convenience when the throat edges are formed in the *brake* rather than the hand machine generally used for this purpose.

The *track method* has become first choice in most *shops* since it requires less labor and what is more important, the *vanes* can be made from heavier metal. This gives the shop an outlet for heavy 22 and 20 gauge scrap material, which ordinarily would never be used.

The **single vanes** for the *track* should be cut 7″ wide, then rolled to a quarter-circle with all edges being raw.

Two pieces of metal comprise the **hollow type vanes,** with both pieces being rolled into a quarter-circle, *A*. Then, by bending back 1/2″ on both edges of one blade, as shown in *B*, the second can be snapped into place, as shown in *C*. To complete the *vane*, the hem is closed tightly in the *brake*, Fig. 8-5.

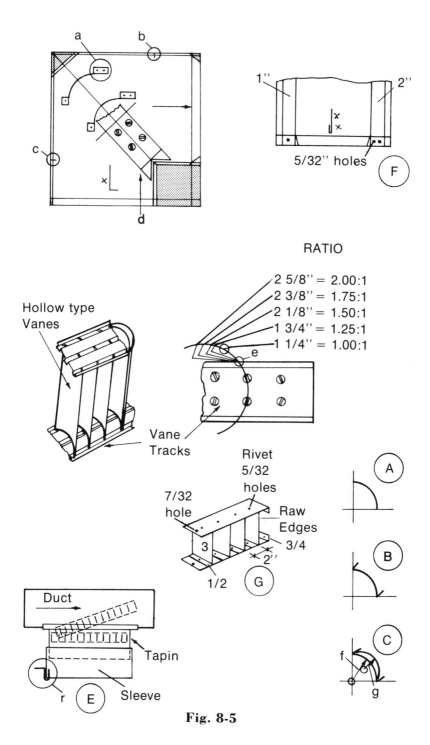

RATIO

2 5/8'' = 2.00:1
2 3/8'' = 1.75:1
2 1/8'' = 1.50:1
1 3/4'' = 1.25:1
1 1/4'' = 1.00:1

Hollow type
Vanes

Vane
Tracks

5/32'' holes

F

7/32
hole

Rivet
5/32
holes

Raw
Edges

3/4

3

2''

1/2   G

A

B

C

f

g

Duct

Tapin

Sleeve

r   E

Fig. 8-5

Duct Layout and Fittings

Another *method* of making the *hollow type vanes*, which is definitely faster and easier to develop, can be handled as follows: take two pieces of metal which are cut the same width, 6-7/8", and run through the *rolls* in the same manner as applied to the other *vanes*. Then, take one piece and put through the *acme machine*, keeping the curved inside of the *blade* up. Both edges having been put through the machine, the other rolled blade, with both edges raw, is slipped into the *acme* edges to complete the double *vane*.

As shown at *f* in Fig. 8-5, the actual radius is 4-1/2", with the outside half of the hollow *vane* having a 2-1/4" radius when completed, *g*.

Since the actual spacing between the *vane* is 3-1/4", a large number of vanes are required in *elbows* of considerable width. Many *shops* omit every other *vane* on *elbows* having very large cheeks.

The *vane* lengths should be cut about 1/8" under the required height. On very large *elbows*, the *vanes* should be divided into two or three stacks, rather than attempting to make them in one long length. Although most *shops* are equipped with a 4-foot *rolls*, it is seldom possible to roll the *vanes* without the center area being flat or shallow. Therefore, it is practical to form *vanes* of this length in the *brake* if such lengths are required.

Fig. 8-5, *E*, shows how **straightening vanes** are used inside; they are also shown at an angle in the form of a **scoop** or **extractor**. As shown in the *blow-up, G*, these are actually the same as *friction dampers* and, being connected with one rivet, they are adjustable. When mounted in the *tapin*, parallel to the *duct*, these *blades* are either used as deflectors, to direct the flow of air through a *grille* or *outlet*, or as a volume control to taper off the amount of air permitted to enter the *tapin*. This actually is a matter of the type *outlet* used; also, the distance the *outlet* is located from the *duct line*, where the *tapin* is mounted. Point *r* indicates the *plaster ground* used to mount *grille*.

When the *elbow* cheeks are of the same width, the *vanes* are mounted on the *track*, which is correct. As shown in Fig. 8-5, this is considered as having a ratio of 1:1 which places the vane equidistant, about 1-1/4", over the edge of the track on both sides, *e*. As the ratio changes, which occurs when opening sizes differ, the *vanes* must be moved around in proportion to the ratio calculated, in the direction of the air flow. The amounts shown in Fig. 8-5 cover practically all ratios the *cutter* may encounter.

As shown in Fig. 8-5, some *cutters* let the *track* carry all the way into the heel, notching out as needed and doing the same with the throat, *d*.

However, this is not necessary; the *track* being metal screwed into place makes a better job. It also strengthens the *fitting* which is important on very large *elbows*.

Fig. 8-5, *F*, shows a *pattern* for the **rivet type blade;** tab allowance being 3/4″ and the side hems 3/8″. The block-out or shear size for the *vane* should be 10-3/4″ x *elbow* height, plus 1-3/8″, which allows 1/8″ under the *elbow* height when formed.

DAMPER DETAIL

Perhaps one of the many items the *cutter* is required to make

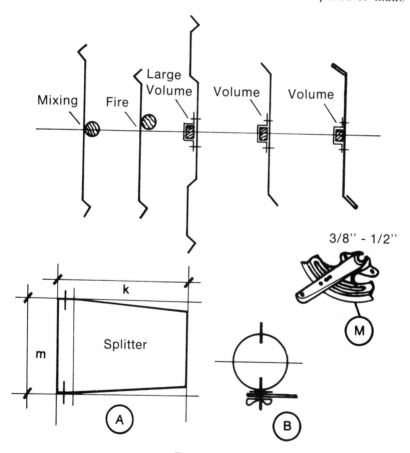

Fig. 8-6

Duct Layout and Fittings

more than any other is *dampers*. Since *dampers* are so simple to make, many mechanics are inclined to overlook the actual care and consideration that should be used in their construction and installation.

A loose or sloppy fitting *damper* can cause a rattle, heard throughout the *system*. Since this will seldom be detected until the job is finished and being balanced, it sets up a real problem. The *damper* causing the trouble is not easy to track down. Then, there is the additional problem of reaching the location of the *damper* in order to make the repair. This is all added labor on the *job*, resulting from someone's carelessness.

Although it may not be given much thought, there are actually

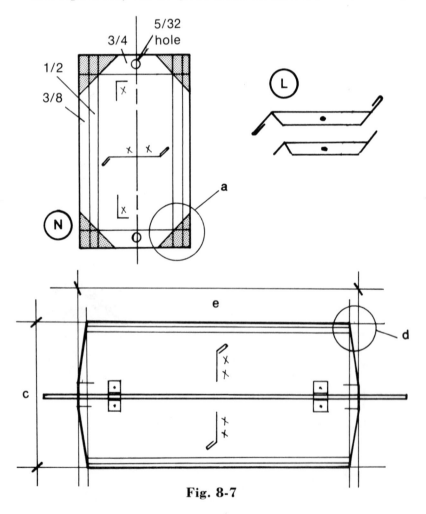

Fig. 8-7

Dampers

six different types of *dampers* used in the design of the average commercial type of *air conditioning system*. This does not include the back-draft *dampers* often used at windows and exhaust *fans*.

These six types of *dampers* are named according to the purpose they serve in the design of the *system*, Fig. 8-6. They are as follows:

1. Volume	3. Splitter	5. Mixing
2. Fire	4. Scoop	6. Friction

Although all these *dampers* can be made in the average *shop*, both the **fire** and **mixing** *dampers* are usually purchased from *firms* who specialize in them. For the busy *shop*, it is seldom considered practical to take the time needed to make them. However, the *cutter* is occasionally required to make them, particularly the *mixing damper*.

Instructions regarding the making of frames, blades and hardware assembly are available through the *distributor* of the *damper hardware*. Therefore, this will not be discussed in detail.

Some shops use *quadrants* and *square rods*, as shown in Figs. 8-6 and 8-7. However, most of them use the regular damper sets, which are purchased, consisting of a square lug on one end with a rod with a spring on the other. These are quickly riveted to the blade.

When making the *volume damper* for unlined *duct*, Fig. 8-7, the

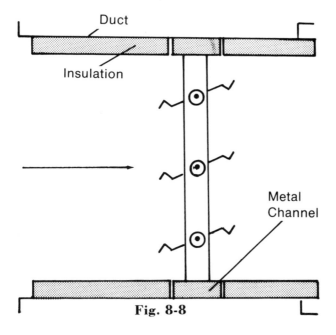

Duct

Insulation

Metal Channel

Fig. 8-8

Duct Layout and Fittings

length e should be the same width as the *duct*, notching the corners off about 1/2'', d. The width c can be about a full 3/8'' under the duct width. This type of *damper* is never fully closed and is set only once, at the time the *system* is balanced.

The lugs for holding the square rod on the *damper blade* can be purchased or made. A reminder regarding square rod: when using it for a *damper* rod, the end of the rod should be notched with a hacksaw, parallel with the *damper* blade, to indicate its direction.

Dampers used in lined duct, Fig. 8-8, when the same length as the duct, have the lining cut out. When the insulation is left intact, the metal bearing plate is used, with the *damper* being made about 1'' shorter than the *duct*.

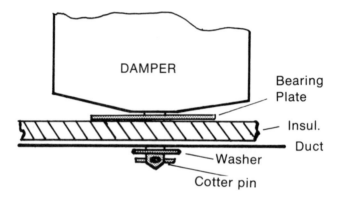

DAMPER

Bearing Plate

Insul.

Duct

Washer

Cotter pin

Fig. 8-9

It is important, when using the bearing plate shown in Fig. 8-9, that certain precautions be taken to prevent the *damper* pin from coming out of the hole in the duct. This is not a difficult problem when using long square rod. However, the regularly purchased *damper* set, which has a spring and round rod, is usually too short. To overcome this problem, place a large washer over the rod and, with a small hole drilled through the rod, insert a cotter pin such as shown in Fig. 8-9.

Particular care should be used in the *layout* and design of *dampers* that are used near the *blower*. They are subject to tremendous pressure. Therefore, to keep them from flapping, make the blades as narrow as possible, using very heavy metal, not under 16 gauge.

It is not practical to use very long blades, especially under the above conditions. Whenever the *author* encounters this situation, he places a partition across the center of the duct. Then, he uses a long 3/8″ rod the full width of the duct, which is locked to the blades with either U bolts or metal clips.

One of the most common errors, when installing a *damper*, is making the hole for the rod too large, which will definitely cause the blade to rattle. To overcome this, drill a smaller hole than the actual required size. Then, using a drift pin from **inside** the duct, gradually make the hole large enough so that the rod can be forced in with the slightest amount of pressure, thereby leaving no room for any excess play around the rod.

The **friction damper,** shown at *N* in Fig. 8-7, can be blocked out with the corners cut off, as indicated by the shaded area *a*. Since they are seldom wide or long, they do not particularly require heavy metal. The edges can be hemmed or left raw, with a slight kink, which is shown at *L*.

This type of *damper* is riveted into place, almost always in a ceiling drip, directly in back of a *grille*. Its purpose is to control the air volume through the *grille*. These are riveted into place and are hand-operated, the *grille* having to be moved in order to reach them.

The **scoop damper** is used where a branch line taps into a trunk line. This is actually a form of *volume damper*, since it must control the air into the branch line. The blade may be made either straight or on a slight radius. This is best shown in the **tapin** section in this *book* (see **Index**).

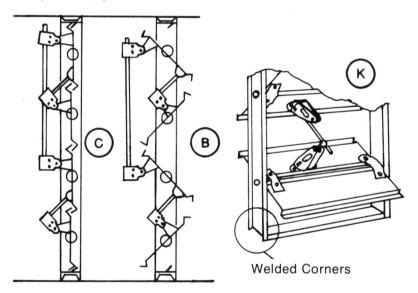

Welded Corners

Fig. 8-10

Duct Layout and Fittings

Fig. 8-6 shows the **splitter damper,** *A*, which is used to control the air flow into two branch lines: usually, where two elbows tie together at the heel. However, this is not always the case. The width *m*, will be the same as the fitting to which the damper is tied. This can be either hinged from the heels with a control rod or use a vertical rod and quadrant. The length is usually about the width of the largest connecting fitting, but can vary either way, *k*.

Fig. 8-6 shows the round type *damper*, *B*, using snap-on end pieces and a specially designed quadrant. The actual *damper* is cut considerably smaller than the pipe it fits into. However, the same theory is applied here: that no *damper* is ever closed off one hundred percent, with the exception of the following; **fire, mixing** and **back draft.** Only these specific *dampers* require a complete block-off.

Fig. 8-10, *K*, shows a cutaway view of a *mixing damper*, partially opened, with the same being shown at *B* and *C*, both opened and closed. These illustrate what is referred to as the **opposed** positions. Being motorized, it is possible to mount these *dampers* in any position.

Fig. 8-11 shows one use for these *dampers*, in what is referred to as the **mixing damper box.** This type installation favors a parallel arrangement such as shown in *O*. This unit is a simple, 3-sided *fitting* with two throats and requires an *access door*. It can also be

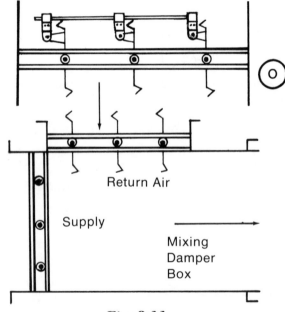

Return Air

Supply

Mixing Damper Box

Fig. 8-11

used as *hot* and *cold* lines, although these are made in a vertical stack with an insulated panel separating the sections.

The *fire* is either controlled by motor or, as most commonly done, is installed with a wire and fusible link. In either case, the blade is pivoted on a bearing located at one-third of the blade's width.

Since the *dampers* offer one of the few opportunities to make use of the heavy scrap in the *shop*, the *cutter* should use 16, 18 or 20 gauge metal. This makes it unnecessary to hem the edges.

REGULAR TAPINS

Shown here are the various types of **tapins** which the *cutter* must be able to make. These also, like the *dampers*, etc., are considered routine *fittings* by the experienced *cutter*, since he makes them almost every day, in one form or another, and usually in multiple lots.

Most *shops* have a certain type they use extensively and the *cutter* is required to make them accordingly.

Disregarding this fact, the *cutter* should be familiar with all the various types he may encounter.

NOTE: For tapin type **access door***, see* **Index.**

The type shown in Fig. 8-12 has become more or less obsolete, as the slight radius which is obtained in the heel does not compensate for the time required in both the *layout* and *fabrication*. Also found to be unsatisfactory, with this particular type *tapin*, is the throat. A certain amount of the *pittsburg* at the throat, as shown at *a*, cannot be made up. It is considered as wasted area, since it lies too close to the *duct*.

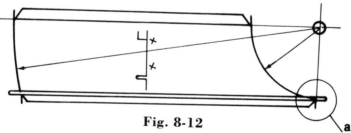

Fig. 8-12

The *tapin* shown in Fig. 8-13 is a definite improvement since this type can be fabricated more easily: also, the squared-off heel makes it possible to *lay out* the *tapin* in 2 parts, unless the size is too large to handle.

To overcome the difficulty found in Fig. 8-12, the throat radius is made 1' under the actual height of the *tapin*, shown at *h* in Fig. 8-13.

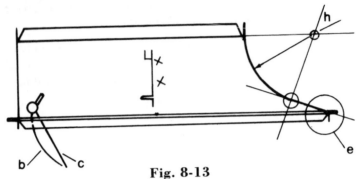

Fig. 8-13

This is then carried to the desired end of the *tapin* opening, with a straight line *e*. In some cases, it is even possible to use a smaller radius and then, by carrying down a straight line on a 45° angle, have a decidedly easier *tapin* to make up.

This makes the actual opening far smaller than required for the other *tapin method*. It is, therefore, a question of whether or not a *tapin* of this type is acceptable.

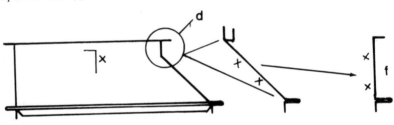

Fig. 8-14

Fig. 8-14 shows one of the easiest *tapins* to make up; the break down, shown at *d*, is 2″. Where a lock is used instead of the *flange*, this 2″ break can be eliminated, making the 45° directly off the lock edge allowance.

All three of the *tapins* shown are usually made six inches high, unless otherwise specified. The *tapin* edge is formed by hand, being 7/8″ + 5/8″ + 3/4″. Many *shops* use a machine for this edge, which requires 1-3/4″ of stock and makes a smaller edge, 5/8″ +

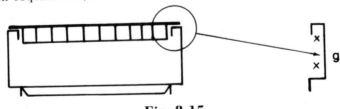

Fig. 8-15

1/2''. The machine is satisfactory for small *tapins* but highly impractical for the larger sizes.

The *tapin* shown in Fig. 8-15, which is used on exposed work, makes a very neat looking job and is both easy to make and install. They are usually kept about 4'' long, but it can vary. The side view, *g*, of this *tapin* shows the difference between it and the conventional type, *f* in Fig. 8-14. Both ends of this *tapin* are kept 1'', with the usual 3/4'' turn under.

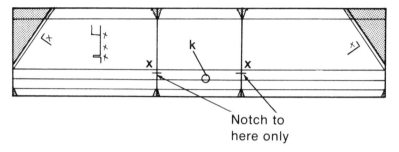

Fig. 8-16

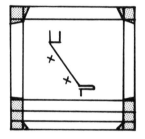

Fig. 8-17

Fig. 8-16 shows the *layout* prepared for the *tapin* described in Fig. 8-14, using a lock as required on supply runs of *duct*. The material is easily blocked out and, by having a scriber for the *tapin* edge, it is a simple routine *fitting*. The 45° ends can be scribed easily, using an adjustable *square* off the lock edge. The end with the *pittsburg* is shown in Fig. 8-17: this also is scribed off the edges of the metal. The lock edge must have a 45° break marked down. However, in most of the *shops*, it is not necessary to mark every break as any experienced *shop* man can easily handle the *fittings*, with very few marks being required.

Fig. 8-18 shows the blocked out *tapin* of Fig. 8-15, with the *layout* as it should appear when notched, as indicated by the shading used on this and other figures.

Duct Layout and Fittings

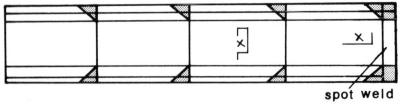

spot weld

Fig. 8-18

As this type is used for mounting a *register* or *outlet*, it will be noticed that a small edge is turned inside for a stiffener, making a solid frame for supporting the outlet.

Place a prick mark on the second line as indicated at *k* in Fig. 8-16; also make straight notches up to the marks indicated by the arrows, keeping back about 3/16'' or 1/4'' from points *x*. This keeps the *fitting* from bending at the points *x*, making it easier to locate the prick marks when forming up in the *brake*.

ROUND TAPINS

The **round tapin collars** shown in Fig. 8-19, *A*, *B*, *C* and *D*, represent the different *methods* that can be used in developing this type of *tapin*.

A has a 1/4'' edge turned out, and a narrow strip of metal about 1-1/2'' wide, being rolled and riveted or spot-welded into place, so that about 3/4'' extends below the turned edge. This is then notched all around to complete the *tapin*. This makes about the best job of all. This same type of *tapin* can be made up by turning the edge and riveting in place the metal strip, which has been run through the automatic *notcher* while in the flat and has the *acme* edge in first. This is then run through the *rolls* which has a slot for just such a purpose. When using the latter *method*, however, it is advisable to turn out a wider edge, 7/16'' if the *rolls* can handle it, and also a wider strip, about 2-1/2''.

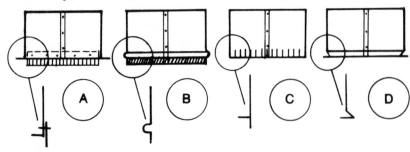

Fig. 8-19

B has the regular *bead* and *crimped edge.*

C has the *dovetail* edge, which consists of straight notches all around at about 1/2″ intervals, depending upon the diameter, and about 5/8″ deep. This continous series of tabs is then turned one in and one out.

D has an edge put on with the *elbow machine*, turning it inside and making it as deep and sharp as possible without cracking or cutting the edge. This type of *tapin* is referred to as the **screw-in**, since it is installed by doing just that. This is an excellent *method*; however, it requires an accurately cut hole in the *duct*. If too large, it will push right up into the *duct* and, if the hole is too small, it will be quite difficult to turn into the hole. The block out for this **screw-in** type is the *circumference* plus one inch for the *acme seam*. Use required height plus 1/4″ for the other size and on one end of this latter size make a 45° notch, using 5/8″ x 5/8″. It will be this edge which is turned in the *elbow* machine.

TAPIN FITTED TO HEEL OF RADIUS ELBOW

This type of *fitting* requires the **tapin** to be on a radius on two sides, such as shown in Fig. 8-20, at *a* and *b*.

These sides will be referred to as cheeks, since they are made **right** and **left**, having single edges, as shown at *C*, similar to regular *elbow* cheeks. The bottom, or shortest length, will be referred to as throat and the longer top as heel. These will be made with the regular conventional type of *tapin* edge, with the throat breaking over 90° and the heel under, as required.

The 1/4″ edge must be turned to the outside as shown in *A*. For the curved *pittsburg* piece, find the length of the curve from points *c* and *d* and shear a piece of metal to that length and about 6″ or 7″ wide, using no heavier material than 26 gauge.

Run this through the *pittsburg* machine. Then mark off the two breaks as follows: To the width *e*, shown in *B*, add enough material to make the width *g* 7/8″. Then add the balance of the usual *tapin* edge which would require another 7/8″ plus 3/4″ of stock, shown at *f*. The surplus material is sheared off and the piece is formed to look similar to the regular *tapin* or *hammer-lock.*

The 3/4″ edge is then run through the *crimper* which will draw the *pittsburg* around to fit the desired radius.

These pieces are then put on the cheeks before the *tapin* is fabri-

Duct Layout and Fittings

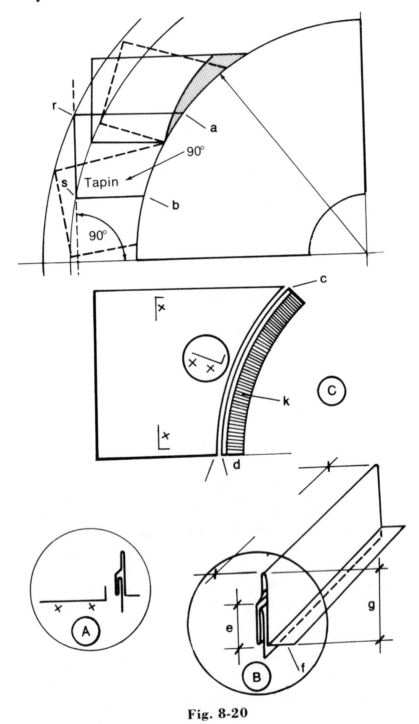

Fig. 8-20

cated. The *tapin* is handled in the same manner as a straight *tapin fitting*.

Although the curve of the *tapin* will fit at any point of the *heel* on the larger *elbow*, a definite fixed location of the *tapin* must be known before attempting the *layout*.

The important point that the *reader* should understand is that, without changing either the *heel* or *throat* of the *tapin, r-a* or *s-b* in Fig. 8-20, there is only this **one location** for the *tapin* along the entire curve of the *elbow heel*.

Be sure that the *tapin* is parallel to the base of the *elbow* as shown in Fig. 8-20. A more definite way of checking is to determine if opening *r-s* is at an angle of 90° to the base, as the slightest move up or down changes this *degree*.

In order to square up the *tapin*, new cheeks are required; also a longer throat **or** heel depending on the direction moved.

ACCESS DOOR, TAPIN TYPE

This type of **access door,** shown in Fig. 8-21, is used on insulated *duct*, with the height *a* being $1''$ or $2''$ high, as required.

The **tapin** edge is identical to the regular *tapins* shown in this Chapter, with the same dimensions and *layout* for this edge applying here also.

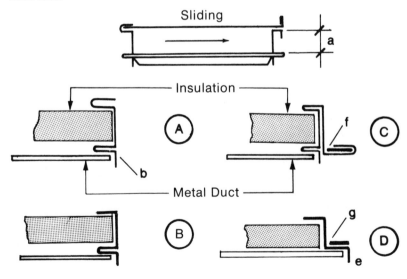

Fig. 8-21

Duct Layout and Fittings

This *door* is of the sliding type and can be made in two different ways, as shown in *A* or *B*. Even though the frame of *B* will be found easier to make up, it is not as practical as the type that is illustrated in *A* since the *door* must wrap around the turned-out edge of the frame in order to slide, having to ride back and forth against the insulation. The alternate *method* has the door sliding directly in the metal itself, which keeps the door from cutting into the insulation.

The hinged type *door*, however, is used considerably more often and is actually more efficient, due to the *felt* edging used to make a seal, *f* and *g* in *C* and *D*.

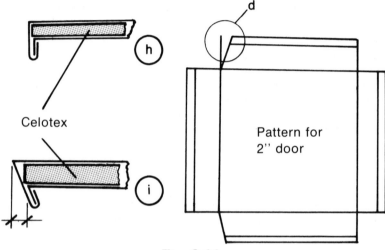

Fig. 8-22

The type of *frame* shown in *C* is similar to the regular *penthouse* door, described elsewhere in this *Chapter*. This *frame* is for a 2″ thick *door* which must be cut back, as shown at *d* in Fig. 8-22, in order to use *hinges* and allow the *door* to swing open. This is often made with regular *sash locks* to keep the *door* in place and whereby the entire door can then be lifted straight out. When the *door* is only one inch high, it can be made as shown in Fig. 8-22, at *h*.

These *doors* are of the regular *tapin* or *hammer-lock* type, *b* and *e*, Fig. 8-21. This type of frame can easily be made in one piece and the two long breaks are then made after making the required notches, similar to *tapins*.

The frame is then easily pulled around by hand to form the desired shape and if a *spotwelder* is available, the door frame is quickly completed.

The frame *layout* shown in Fig. 8-23 is for the frame of *B*. To make

212

a similar *layout* for the frame of *A*, the shaded area shown across the top is omitted and then formed across the entire pattern, as is indicated on the *pattern* in Fig. 8-23.

The *reader's* attention is called to the fact that these *doors* are usually only about 8″ to 12″ square, as they are usually used as a hand access to adjust a *damper* or to replace a *fusible link*.

Shown in *E* is a *blow-up* of the notch required in the Fig. 8-23 *layout*, which is typical of that used in the regular *tapin*.

When an insulated *door* is required, it is necessary to make the frame in much the same manner as the *penthouse type door*. For its construction, the *reader* is referred to **penthouse door**, included in this Chapter.

Two different frame elevations are shown: one is for one-inch insulation and the other for two inches, which is the outside thickness or the *plenum* or *duct* insulation in which the *door* is to be installed.

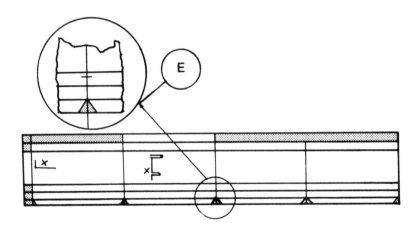

Fig. 8-23

PENTHOUSE DOOR, INSULATED

This **door** is the filled or insulated type in which 1/2″ Celotex is sandwiched between a double wrap of metal, as shown at *d* in Fig. 8-24. A corner detail of both panels of the *door* is shown at *m* and *n* with the correct notching. Shown at *E* and *F* is the elevation of these same panels as they should be formed.

The *frame* shown in Fig. 8-25 is usually made 24″ wide by 78″

213

Duct Layout and Fittings

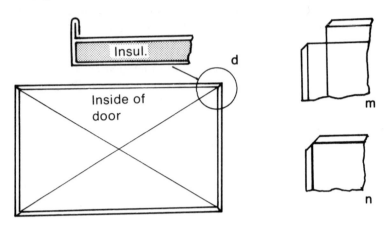

Fig. 8-24

high and is installed 6″ up from the floor. The *door* is the *tapin* type, shown at *A*, which is an end view of the *frame* for an uninsulated *housing*. For insulated *housings*, an extended frame is used and this is shown at *B*. The construction of both *door* and *frame* is quite simple; however, it must be carefully laid out and formed on the *brake* with equal care.

R, in Fig. 8-25, shows an enlarged view of the *frame* corner *C*. Shown at *H* is a *frame* which requires 6-1/4″ stock.

For the extended *frame*, use 11-1/2″ stock if the insulation is 1″ thick. These figures, as shown in *H*, provide a satisfactory *tapin frame*; however, some *shops* have preferred requirements as well as patterns on hand.

The *door* should fit the *frame* with less than 1/8″ clearance all around.

As shown in Fig. 8-25, *E* and *F*, the outside thickness of the *door* is 7/8″, the inside section having a 5/16″ edge turned up. Therefore, with the Celotex being 1/2″ thick, there is 1/16″ play between the two sections. The foldover edge can be 3/8″; it will lose just enough in the forming process to make a snug fit when finished, as it must be completed by hand.

It is also important to panel or cross-break both sections, so that the rise of the cross-break is towards the Celotex or the inside.

For best results, use 24 gauge metal for the *framing* and 26 gauge for the two *door sections*.

Whenever possible, the *door frame* can be built directly in the

214

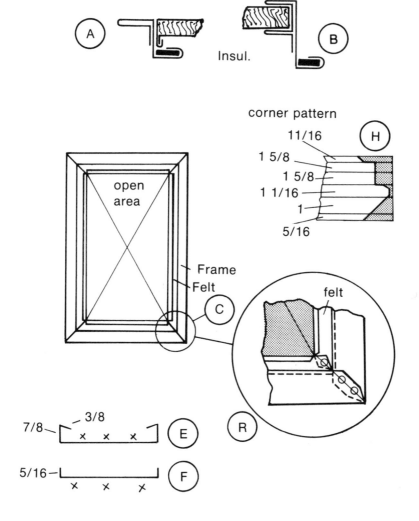

Fig. 8-25

plenum or *housing* wall, as shown in Fig. 8-26. This assures less work in the fabrication and greater strength of the *fitting*. It is necessary to turn the standing seam over and above the *door* to the inside. By doing this, the single edge carries through with the *frame*, as it is not necessary to notch out at the four corners of *door frame* where the seam continues on. *G* shows one of the four corners where this occurs. Regular 3″ x 3″ *butt hinges* are usually used. For best results, use 3 *hinges*.

Duct Layout and Fittings

As for the *lock*, no definite style is required. However, as a safety factor, *handles* must be on both inside and out, so that anyone can operate the *lock* from either side.

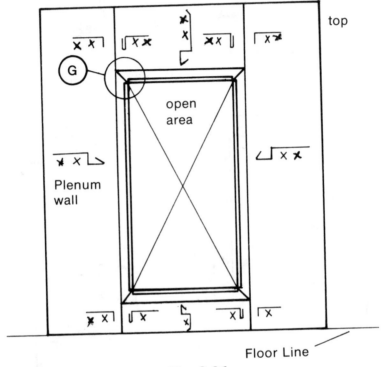

Fig. 8-26

SLIDING DOOR DETAIL

This type of **access door** is used when an opening is required in a particularly crowded area where space will not permit the use of the conventional type *hinged door*.

The rectangle *A, B, C, D*, shown in Fig. 8-27, can represent the side of a *plenum* or joint of *duct*. It is merely to show the actual detail of the *door* and slide rails, as they appear when placed in position.

The dotted line indicates the actual hole cutout, with the other dotted lines being the parts of the frame or slides which are inside.

Shown at Fig. 8-27, at *G*, is the *layout* in the flat of the side rails of which two are required, making a right and left. The width of *f*, *g*, and *k* should be 1", with *e* being either 3/8" or 1/2" for a hem. Have the width of *h* and *j* total about 1-1/2" together, putting a

216

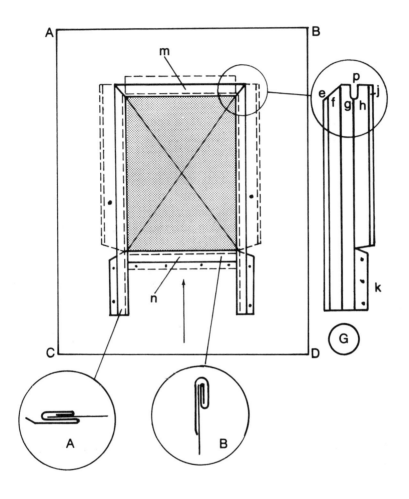

Fig. 8-27

slight kink or feathered edge at *j*, thereby making a smooth and snug fit. The end *elevation* of these rails is shown in *a*, with the light line indicating side of the *door*.

The corners are at 45°, as shown in Fig. 8-28, with the slot shown at *p* reaching down to the same depth. This corner *layout* also applies to both ends of the top frame *m*, shown in Fig. 8-27. It is important that this *frame* be set in place first, since the side *frames* hold it in position.

The bottom *frame*, *n* in Fig. 8-27, is cut off at both ends square

Duct Layout and Fittings

and an *elevation* of this is shown at B, with the light line representing the *duct* or *plenum* to which it is riveted. The dots shown in Fig. 8-27 are rivet locations. The actual proportioning given frame section n is not too important, since it is merely a double-hemmed edge on which the *door* rides. It also forms a closed edge when the *door* is in the closed position.

The *door* itself should be made of heavier metal than the *frame*, with the corners notched out as shown in D, Fig. 8-28. Also rivet a small piece of angle iron on the notched end. C shows how the angle should be riveted into place on the *door*.

The actual length of the *door* is determined by just how much edge is hemmed over on the frame section n, Fig. 8-27. The length of the door must be closely measured, as the angle must extend beyond this part of the frame, allowing just enough to close at the other end; at the same time, it must fit up snugly against the double-hemmed edge of frame section n.

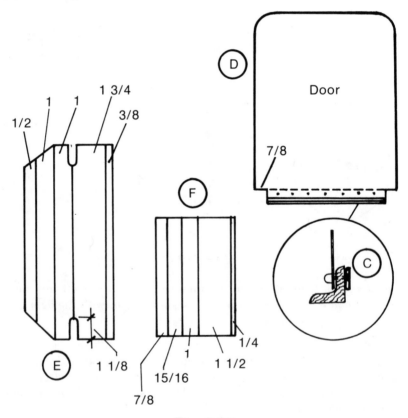

Fig. 8-28

218

E, Fig. 8-28, is the *pattern* as it should be made for the top part of the *frame m*, *F* is the *pattern* for *n*, which is the last fitted part of the *frame* and is riveted to hold in position.

Cut the *door* about 1/4″ under the width between the side slide rails. Then use a file to round off the two corners which enter the side rails. Also file the side edges to make certain that the *door* will not bind. If necessary, apply a few drops of oil or apply a light coat of grease to the sides. Although the *door* is required to slide easily, care must be taken to have it tight enough so that vibration will not cause the *door* to work its way out.

DOOR DETAIL

SIGN   Although very satisfactory on a *sign*, this design is not suitable for *duct systems* with air under pressure. It can be used, however, as a temporary patch, when sealed with tape. Use fairly heavy material for the *door*, about 1″ larger all around. Fig. 8-29 shows the door, set in position over the opening, with *K* being an enlarged view of the cut on top. The offset required for the *door* is shown at *M*. Use a pair of metal screws to hold in place.

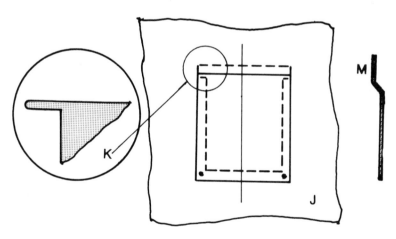

Fig. 8-29

PATCH   This type of door, Fig. 8-30, is probably the most commonly used *method* for covering an access opening. Since it can be made leak-proof by gluing *felt* or *rubber* strips around the edge, as shown in *L*, it can be used when under pressure. Shown in *P* is the inside *plan view* of the *door*, with *N* showing a corner *pattern* and *y* being the cut-off. Metal screws or *sash locks* are used to hold *door*.

Duct Layout and Fittings

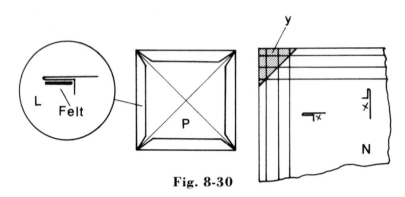

Fig. 8-30

JOINING TWO FITTINGS TOGETHER

It is often necessary to **join two fittings together,** such as two *elbows* at the heel, *r* and *s* in Fig. 8-31.

When this is required on the flange end of the elbows, it is necessary to cut back both heels in order to keep clear of the lock into which the flange will be placed. This is shown at *a*, with a *blow-up* section shown also in *F*. The raw edge, *b*, is cut back 1-3/16,'' with the hook edge, *c*, back 1-1/8''.

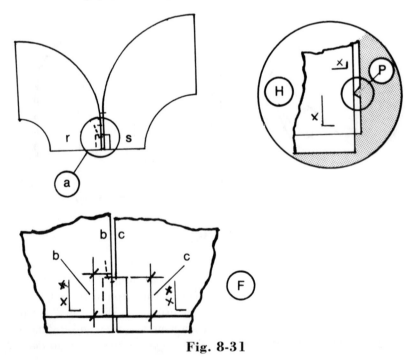

Fig. 8-31

Joined Fittings

The single edge of the cheek is left on the *fittings* and merely flattened out as shown. This makes a closed connection of all pieces concerned. A better example of this notching of the cheek is shown in *H*, with the cut of the small *vee* indicated at *p*, being up 1-3/16″ off the bottom or flange end.

When joining *fittings* at the lock end, it is not necessary to make this deep cut-back. In fact, many *cutters* make no allowance here as the lock is above the actual *fittings*. This is quite true and, in some cases, proves to be satisfactory. However, in most cases, the thickness of the metal from the *pittsburg* locks will keep these *fittings* from drawing up tight and still be kept square.

Therefore, it will be found best to cut the raw edge back about 1/2″, to allow for the surplus material.

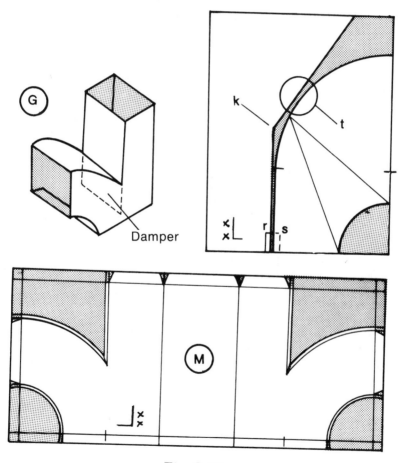

Fig. 8-32

Duct Layout and Fittings

After the connections are all made, the upper part of the cheeks are also held in place with a short piece of hanger strap and metal screws. The *fittings* must be kept securely squared off when making this final connection.

The joining of two *fittings* is not confined to *elbow* heels, since numerous combinations are used to meet the conditions on the job.

In making a similar type of *layout*, as shown in Fig. 8-32, it is necessary to use care in the *layout* of the *transition fitting*. When only a limited amount of space can be allowed, as at point t between the two *fittings*, it makes the locating of point k very important. It is usually best to make up the *elbow* cheek first. This will serve as a guide to establishing point k.

Attention is also called to the openings r and s of these *fittings*. Whenever two *fittings* are placed together in this manner, there is a certain growth which must be allowed for, due to the thickness of metal and the *pittsburg locks*.

In order to compensate for this growth, which will vary according to the gauge of metal being used, figure that 1/8'' for each *fitting* of 26 or 24 gauge will be sufficient. However, it would require at least 1/16'' more per *fitting* when using 22 or 20 gauge metal. Much will depend upon how carefully the *fittings* are fabricated, as this is something that the *cutter* cannot control or make allowances for. It is often better to make the total allowance from one *fitting*, such as when an *elbow* is to be tied in with a piece of *duct* or some straight *fitting*. By taking the full allowance from the *elbow*, no change is necessary in the development of the other *fitting*.

It is often required that a narrow piece of duct, possibly just 2 or 3 inches deep, be tied in with an *elbow*. Here, also, it is practical to make the total deduction from the *elbow*.

The one piece *fitting* shown in G serves a purpose similar to the joining of separate *fittings*. When it is permissible to use this type of *fitting*, considerable time and labor is saved. The simplicity of the *layout*, shown in M, makes this fact quite obvious. The two *wrappers* and *damper* are easily made and fabricated. The joined *fittings* usually have a splitter *damper* in the preceding stub or joint of pipe.

COMPANION FLANGES AND PIPE COUPLINGS
Flat Ring Type Flanges

This type of **companion ring** is made from the heavy scrap metal which is usually a loss to most *shops*. Unlike the regular type of **companion flange,** which is made from *angle iron* and riveted to the duct or *fitting*, this type is loose on the pipe, being kept on by the edge which is turned out on the pipe.

Although it is not meant to serve as a substitute for the *angle iron* flange, it can be used to good advantage on round pipe such as encountered in the high pressure *air conditioning systems*.

It is usual practice in most *shops*, when doing a high pressure *round pipe job*, to put together all of the straight *pipe* and *tapers* into continuous long lengths. By doing this in the *shop*, it is possible to solder all of the joints, to insure air-tight connections.

This leaves only the ends to be made up when installed at the job site. The best means of handling this problem is to make the flange rings of heavy metal, about 1-1/2" wide for a tight and neat looking joint. See *r* in Fig. 8-33. Also shown at *s*, is the same flange when cut square. This is not as satisfactory, as there is too large an area which will have a tendency to buckle when drawn up by the bolts. It is possible to cut off the corners, making a hexagon, and place the bolts closer to the pipe. Also, it is possible to turn out a 3/8" or 1/2" edge all the way around as a stiffening measure.

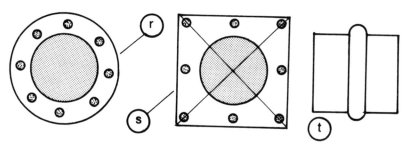

Fig. 8-33

Either *method* will involve more handling and that, of course, means more time is taken up, which only points out the fact that the ring shown at *r*, Fig. 8-33, is the best *method*.

One *flange* is required for each pipe end and, if care is taken to keep the bolt holes for each ring the same, it will not be necessary to maintain matched pairs. To make air-tight connections at these *flanges*, some type of a good *mastic* or *cement*, which will not dry out, must be applied before bolting together.

It is only necessary to use 1/4" bolts, which can be obtained with

Duct Layout and Fittings

pan heads, making it unnecessary to use any washers under the head. It is also a good practice to extend hangers from these *flanges*, using either one or two hangers according to the size of the pipe.

NOTE: Most shops have a circle cutting machine which makes developing the ring flanges a simple process. It is usually possible to cut all the flanges at one time and, as they keep dropping in size after each taper, more than one can be cut from the same piece of metal by starting with the larger rings and working on down to the best advantage.

COUPLING FOR FIELD JOINTS

Although the *coupling* shown at *t* in Fig. 8-33 is hardly classified as being a *companion flange*, it does serve the same purpose. It is a highly recommended *method* of making field joints without using the conventional type of *companion flanges*. It also eliminates the need for *mastic* or *cement*, since the joints can be sealed with *tape*.

The only real objection to this particular type of connection is the fact that it requires careful calculating in regards to having *coupling* fit into the *pipe* without being loose or being so tight that the connection cannot be made. If the pipe being used on the *job* is available in the *shop*, it is possible to work out a perfect fit, even if it takes the trial-and-error method. The *couplings* are made anywhere from 2" to 8" in length, with as large a *bead* as it is possible to put in the center of the coupling.

Understandably, it is not desirable to make any *crimp* or drawing-in of the *coupling*, otherwise the conventional type of *crimp* and *bead* would be used in the first place. However, it has been the *author's* practice with this *coupling*, to make as close a *fit* as possible and, instead of trying to draw in the edge of the *coupling*, to reverse the procedure and put a slight flaring out on the pipe into which the *coupling* is to fit. This works out exceptionally well, as the slight flare fits against the *bead*; and since the seams are taped up anyway, no one is aware of how the joint was actually made.

DRAW-BANDS

Quite often the *cutter* is required to make **draw-bands.**

There are several types used, both **round** and **square shaped.** However, they are quite similar in construction, although some are best made in two pieces, rather than in one wrap-around. This is left to the *cutter's* discretion.

224

Fig. 8-34 shows a typical *round draw-band*, which is usually made from hanger strap or lightweight band iron (flat bar), and is made for a large round-type *canvas connection*, such as shown in Fig. 8-35.

For this type of *connection*, both ends of the pipe are double-hemmed, with a wide piece of *canvas* being wrapped around the pipe enclosing the open area. Allow about 2" or 3" of *canvas* for either sewing or stapling the join of the *material*. The *draw-bands* are then placed over the *canvas*, directly in back of the hemmed edges, and drawn up tightly.

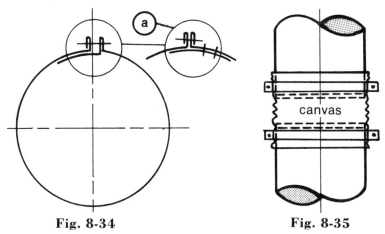

Fig. 8-34   **Fig. 8-35**

A minimum gap of about 1/2" is usually sufficient for the take-up. The connection could be made in one piece with the surplus stock of about 3" being kept down one metal thickness to slide under the opposite end, shown in Fig. 8-34; or by adding a separate piece, as shown in circle *a*, which can be riveted in place, making a very neat job.

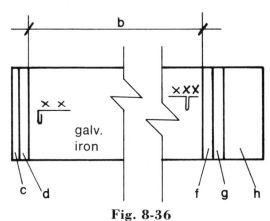

Fig. 8-36

Duct Layout and Fittings

When rolled into shape, it actually forms a *stub* of *pipe* or may be referred to as a *sleeve*. The indicated *d* and *f*, in Fig. 8-36, are made 1″, with *g* being one metal thickness longer than 1″. The hem, *c*, can be 1/2″, with the lap, *h*, about 3″.

This *draw-band* forms the same as that shown in Fig. 8-34; however, the fabrication can be difficult, especially if heavy metal is used.

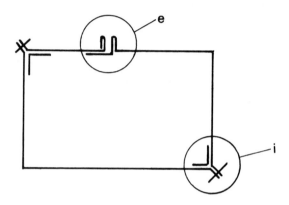

Fig. 8-37

The square and rectangular *draw-band*, shown in Fig. 8-37, can be made with the joints on the corners, *i*, or through the middle as shown at *e*. Basically, the same figures and principles are used here as for the round *draw-bands*.

The added pieces at the corners, *i*, do not have to be riveted in place, as they usually stay there once the *band* has been set.

Here, too, is another of the small items that require care in its construction. A *draw-band* made with too large or too small a gap not only will probably prove inefficient but will make a sloppy-looking job as well.

CANVAS CONNECTION FROM READY-MADE STOCK

Ready-made material undoubtedly is the fastest method of making up the **canvas connection** and many *shops* use it exclusively. It is available in many widths, 9″ is standard, of various gauge metals and types of materials. Most *shops* only stock the 9″ width. When wider stock is required, added metal is seamed onto either end. Another *method* is to add short stubs of pipe having an S lock formed on one end to fit on the raw edge of the *canvas connection*. The, seaming *method* makes the best looking job and, if the *shop* has an *acme machine*, this is a very simple process.

Canvas Connections

Although ready-made material is a faster *method*, many *cutters* find it rather awkward to handle. This is due, of course, to the roll of the material and the thickness of metal being used in the wrapping of the *canvas*.

Here is an easy *method* of making the connection, using **ready-made** material.

STEP 1. Determine length of material required from the given size on the *work sheet*, add to this amount 2", then cut off.

STEP 2. Place on the bench with the *canvas* down and the metal folded in the same manner it came off the *roll*.

At one end, square off 1" from the end. Then measure off the amount of each side, which will leave the same lap on the other end. At each mark along the material, extend the line across to the other edge; then place 3 prick marks on each of these lines. The purpose of this is to mark the underneath piece of metal at the same points, Fig. 8-38.

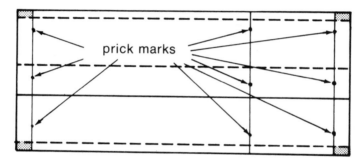

prick marks

Fig. 8-38

STEP 3. Remove each corner of the metal, being careful not to cut the *canvas* outside of the wrapped edge. Use a sharp *chisel* to flatten this heavy edge enough so that the *shears* will handle it with little difficulty. Then, opening this notched piece, lift out the *canvas* and cut out the two corners on one end; on the other end, cut lap clear away, leaving only 1" lap of *canvas* on both ends.

STEP 4. Open the folded metal and place on the bench, wrapped edge up. This will cause the material to be curved up off the bench. To flatten out, use the side of the hammer along the wrapped edge of the metal.

STEP 5. From this point, follow the usual procedure of scribing and notching to complete *layout*.

Duct Layout and Fittings

*NOTE: The method shown here is just one of the many ways that the canvas connection can be made up using the **ready-made** stock. Each shop has its own particular way of wanting it made and the cutter has to go along with that method.*

The *seam* can be made at the corner or, as some prefer, in the center of a *side*, which does have its advantages, such as providing a tight connection and a very neat job. Others prefer to use *cement* to *seal* the *canvas* joints, while still others use the *rivet machine* on both the *canvas* and the metal. However, one point upon which most all agree is to keep the *roll* or bulk of the *canvas* on the inside. When used with the other side in, it is true that the *canvas* is out of the *air stream;* however, this exposes the metal wrapping which holds the *canvas*, and this is apt to cause some turbulence in the *duct line.*

CANVAS CONNECTION, HANDMADE

Although it is generally referred to as a **canvas connection,** other types of material are often used, such as *asbestos cloth* (also available with woven wire), or *neoprene*, which is acid resistant. In some cases, *job* specifications call for a double layer of *canvas*. However, the same procedure can be used in the *layout* and fabrication.

The *flex* or *silent connection*, as it is also called, can be made with a variety of connection ends, such as a flange, lock, hem, cuplock, hammerlock, etc.

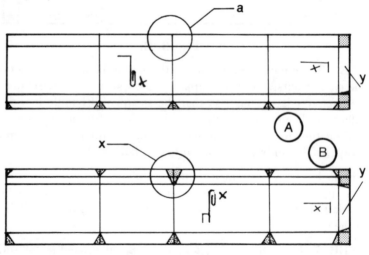

Fig. 8-39

228

Canvas Connections

From the raw materials, make the connections shown in Fig. 8-39. *A* and *B* are the layouts for both sides of a canvas connection, having lock and flange, made in a one-piece wrap-around. Although the size of the connection determines, to some extent, how many pieces are required in its construction, the *cutter* must use his own discretion.

Determine the block-out size and cut material at the *shear*. The length is equal to the perimeter plus 3/4" lap, *y*, Fig. 8-39. The width of the pieces is determined by the required length of the *connection*, plus type of edges required.

Make an allowance of 7/8" for wrapping the *canvas*, this being divided as 3/8" and 1/2". Scribers are used for all marking along the edges of the blocked-out material. The 3/8" edge is notched, as indicated, for most of the light metal; however, for heavy metal, make this notch the full 7/8" as shown at *x*. The flange end takes a straight 7/8" cut as shown at *a*, with the tabs shown at *y* being 3/4" for spot-welding or rivets.

The *canvas* should be cut about 2" or 3" longer than the overall stretchout of metal, in order to have enough material to either staple, sew, or cement the ends together.

The *canvas* must be 3" wider than the desired finished size, as the wrapping process takes up this much material.

The *method* of wrapping the *canvas* is shown at *b*, *c*, and *d* in Fig. 8-40. With the first step *b*, the 3/8" edge is carried over the full

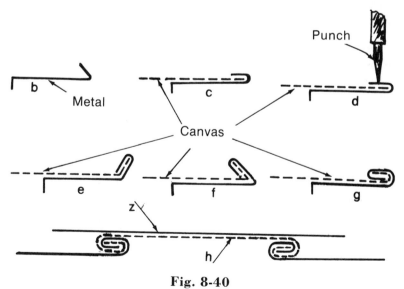

Fig. 8-40

229

Duct Layout and Fittings

throw of the *brake*. The *canvas* is then inserted, and the metal flattened down, *c* and *d*. Using a center or prick punch at about 6″ centers, prick mark the edge to keep the *canvas* from pulling out. Then, as shown in *e*, *f*, and *g*, place the connection back in the *brake* on the 1/2″ mark; or hold even with the outside of the *leaf*, which is the same, throw back all the way and then flatten.

Follow through in the same way on the other side to complete the wrappings of the *canvas*.

Break the ends as required; for the tabs, *y*, use a *tongs*. The other breaks can be made over the edge of a bench by hand.

When making the *round connections*, place a piece of metal, *z*, on the inside against the *canvas*, *h* in Fig. 8-40, rolling this together to the required size. This will keep the *canvas* from cutting as it passes through the *rolls*; however, keep *canvas* taut during the process.

BELT GUARD DETAIL

Belt guards are made in many various shapes in order to meet the conditions at hand. Job specifications almost always include the particular kind of material that is to be used, as well as the type of *guard* required.

In some cases, both the front and back of the *belts* and *sheaves* must be covered, which means that the *guard* has to be made up in sections or possibly hinged.

No attempt will be made to show these various types of *guards*. Shown here are two of the commonly used designs which the *cutter* is often required to make. The ability to make these, along with the ingenuity of the *cutter* himself, should enable him to construct any type of *belt guard*.

The dimensions usually given with the detail of the required *belt guard* are *x*, *y* or *y-y* and *z* in Fig. 8-41; also the diameter of the sheaves and the depth of the *guard*, *b*, with the amount of clearance required. A certain amount of take-up is necessary, since all *belts* stretch when in use.

Using the given dimensions, draw a full-scale outline of the *guard* and from this *sketch*, measure the *stretchout*. Then shear the metal required to make the frame for the outside.

For the **expanded metal** *guard*, the edge should be turned in

230

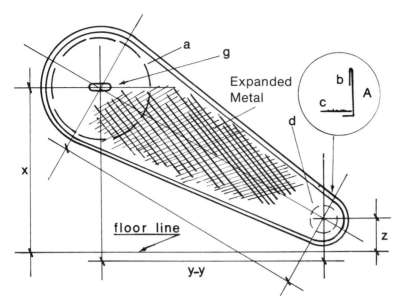

Expanded
Metal

floor line

y-y

Fig. 8-41

about 1/2'', with the expanded metal either spot-welded or soldered in place. The *guard* shown in Fig. 8-42, using **galvanized iron,** has a 3/16'' edge turned inside, with the outside frame having a *pittsburg* completely around it, *B*. *C* shows the crease put through the center for stiffening purposes.

Fig. 8-41, at *g*, shows the slot usually required for using the *tachometer*. To eliminate the raw edges of the expanded metal, make an oval-shaped metal ring and then solder it into place. For the side frame, the one or two short seams required can be riveted.

The type of *guard* shown in Fig. 8-41 usually has a heavy gauge frame, which makes the rounded ends difficult to form. However, in some cases, the 1/2'' edge can be pulled into shape by using the *crimper*. Should this be too hard to manage, as it would be with very heavy metal, notch the edge which forms the half-circles, using care to *vee* out as little as possible, making a neat job. This can then be soldered and filed smooth.

The broken line circles shown at *a* and *d* are the *sheaves*, with *e* and *f* being galvanized iron and *c* being expanded metal.

An easier, faster *method*, using straight lines to replace the half-circles, is shown in Fig. 8-42 by the added lines together with the shaded area. This same *method* is shown when the *guard* is the type that sits on the floor.

Duct Layout and Fittings

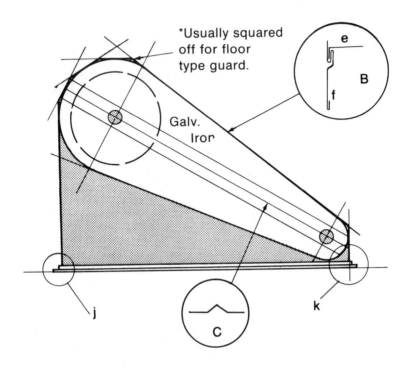

*Usually squared off for floor type guard.

Galv. Iron

e

B

f

j

C

k

Fig. 8-42

When made to sit on the floor, use *angle iron* around the bottom as a means of anchoring the *guard* into place, as shown at *j* and *k*.

DETERMINING ANGLES ON HOODS OR CANOPIES

It is important, when laying out a **hood** or **canopy,** to indicate the amount of *angle* required when breaking up the various *seams, gutter,* and *body section.*

Although this can be shown by *degrees,* that *method* is not recommended, as the margin of error is too great. Since most *hoods* are made of heavy iron or stainless steel, and practically always exposed, it is necessary to eliminate as many chances of error as possible.

By making up *templates* of the various *angles* directly from the *layout* of the *hood,* the correct amount of *angle* can be checked by the *brakeman* quite easily.

Fig. 8-43 shows a **plan** view of a *hood,* to be used as an example in explaining the *method* of procedure in obtaining the correct *angles* for the *templates*. The *method* shown here can be applied to any *hood* or *canopy*, regardless of its size.

STEP 1. Develop the *miter lines*, *u-v* and *w-x*, using dimensions given in the *plan* view, Fig. 8-43. Sections *M*, *N* and *O* do not necessarily have to be drawn out fully; however, since the *templates* are usually made at the time of the actual *layout*, each is completely developed.

STEP 2. *Templates A* and *B* are the correct *angles* for the standing seams of the diagonal *miter lines u-v* and *w-x*. The actual finished height of the *hood* is placed at a 90° *angle* from any of the four corners, *u*, *v*, *w*, *x* in Fig. 8-43.

A connecting line which forms a *triangle*, Fig. 8-43, will give the

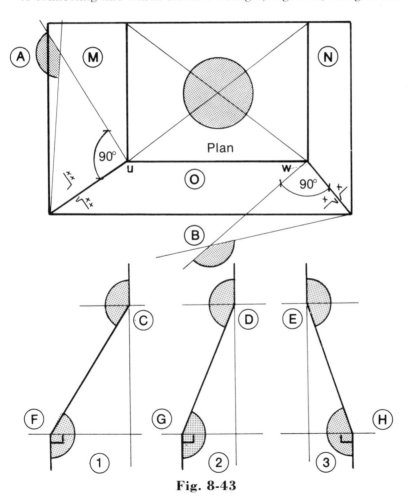

Fig. 8-43

Duct Layout and Fittings

correct angle shown at *A* and *B*. This *angle* will only be for the *miter lines*.

STEP 3. The *templates* for *1, 2*, and *3* of Fig. 8-43 are developed from the actual profile or *elevations* as shown.

The same *template* serves for both the top and bottom breaks as indicated, *C-F, D-G* and *E-H*.

Since it is necessary to obtain the correct lengths through *triangulation* in order to develop the *hood*, no extra, unnecessary *layout* is required in making these *templates*.

The close relationship between the *layout* of a *hood* and the *straight transition* is quite obvious. It is for this reason that actual discussion of its *layout* has been purposely omitted. The *reader* should not encounter any problem here if he applies the same *method* that was given in the chapter on transitions elsewhere in this *book* (see **Index**).

COIL DRAIN PAN

The **pan** shown in Fig. 8-44 is made from one piece of metal, having the drain off the end.

As shown at *b*, the *floor flange* which is used as the connection *fitting* must be kept close to the bottom of the *pan*.

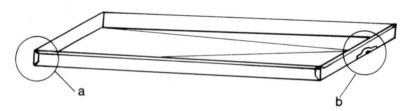

Fig. 8-44

To do this, the *flange* must be cut down as shown in *B*, Fig. 8-46, which allows the outlet to be directly at the bottom of the *pan* without extending below the *pan* itself.

Fig. 8-45 shows a drain *pan* having the *floor flange* directly off the bottom. The round *flange* shown in *A*, Fig. 8-46, is the type usually used for this purpose. Although the drain is shown off the end, *c*, it is often placed in the center of the *pan*. The location is based on the purpose of the *pan* and how much water must be carried off.

234

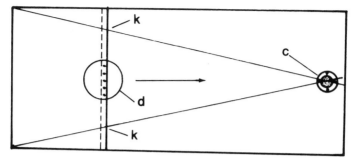

Fig. 8-45

Very often the length of the *pan* is too long to make in one piece, in which case an added piece is riveted and soldered in place. It is important to have the lap of the added piece on top in the direction of the drain, as shown at *d* in Fig. 8-45.

Although no great strain is on this riveted seam, the rivet holes should be about 1″ on centers, using about a 2 lb. rivet.

The purpose of this close rivet *layout* is to help keep the *pan* from buckling when soldering the seam. It is a good practice to place the rivet holes directly on the lines, allowing lap of about 1/2″ on both pieces. If the *pan* must be kept flat, one side of the seam must be offset.

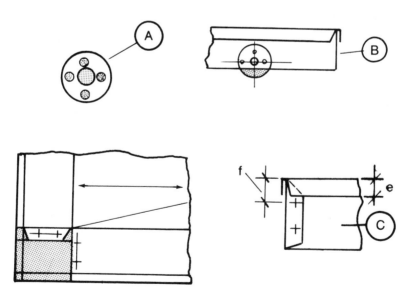

Fig. 8-46

Duct Layout and Fittings

The corners can either be riveted or spot-welded, then soldered. It is important, when the corners are to be spot-welded, that the tabs are left on the short side of the *pan*, as shown in Fig. 8-46 with the arrow indicating the length of the *pan*. Allow about 5/8″ for corner lap as shown; the shaded area is to be cut away.

The *pan* is usually panelled from the corners to the center point of the drain. When long, two-piece *pans* are made, it is necessary to panel each piece separately before riveting together. Make the panel match as shown here in Fig. 8-45 at *k*.

Extreme care should be taken when soldering to insure a water-tight job. The corner rivet holes, shown in *C*, should be kept down about 5/8″, *f*, with the hem, *e*, to be 1/2″. The lap on the corners are to be on the outside, *a*, Fig. 8-44.

Cut the hole for the *floor flange* so that it is about 1/4″ larger all around the *pipe* opening. In this way, the solder will be seen and a positive seal is assured. Make certain the *flange* is *tinned* before riveting it onto the *pan*. The *flange* must be secured to withstand the pull of a *pipe wrench*, which every *fitter* uses when making the tie to the *pan*, with his pipe line for the drain.

The *pan* is not a difficult item to make. However, it presents a challenge to the *mechanic* to turn out a neat, clean-cut job, especially when having to use heavy iron or copper.

ANGLE IRON CONSTRUCTION

The *cutter* should become thoroughly familiar with the use of **angle iron** in developing the many items necessary to the *air conditioning system*. A considerable amount is used in the course of installing the job, for stiffening purposes, hanging *equipment* or *duct* and also for the frame-work for installing the panels which make up a *housing* or *plenum* chamber.

The *shop* man is required to develop a number of different items such as various stands to support equipment, brackets, companion rings and many other types of frames. All of these must be made to given measurements, that are to be kept exact, and to a specific pattern. To *lay out angle iron* under these conditions requires considerable planning ahead and careful cutting of the material. The size of *angle* most often used is 1-1/2″ x 1-1/2″ x 1/8″ stock; but for stands to support heavy equipment, it is usually requested that 2″ x 2″ x 1/4″ be used.

Being able to both *weld* and handle the *cutting torch* is a definite advantage, since about every item made, with a few exceptions, is *welded* together. It is the practice with most *shops* to have the *cutter* prepare or set-up the work for the *welder*. Care must be taken in keeping the work both square and to the required size. No standard procedure can be given as to how the various items should be made. However, it is possible to use one basic step in developing most of the different items, as shown in Fig. 8-47 at *m*.

This notch is made with a *vee notching machine*, which is used in most *shops*. The notch makes a 45° cut on each side, so that a 90° corner is made when pulled around square, as shown in the bracket at *n*, and which also is typical of all corners of the wrap-around *frame* of *P*.

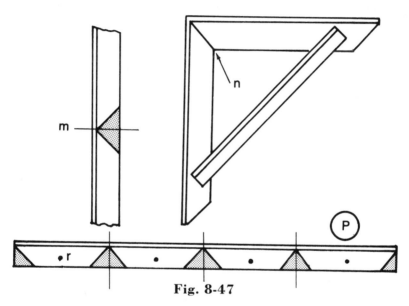

Fig. 8-47

When making the *layout*, allow for the thickness of material if an outside dimension is required. This machine can also be used for single cut-offs, when the frame size makes it necessary to build in 4 pieces or possible halves.

As shown in *P*, by making these notches 4″ on center, these pieces can then be cut apart and used as feet for *angle iron legs* to carry the *duct work* across a roof. It is the best practice to either drill or punch holes for the bolts, *r*, before cutting apart. Some *shops* prefer to use 4″ square heavy metal pads, in place of this angle, which are welded on the angle.

The *angle iron frame* shown in Fig. 8-48 is used as a *companion*

237

Duct Layout and Fittings

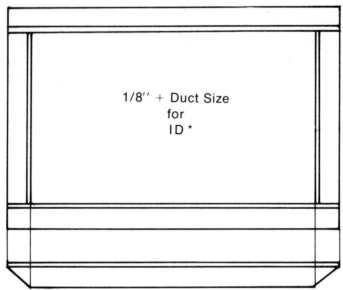

1/8″ + Duct Size
for
ID *

*When metal thickness of duct is extra, this should then be 1/8″
+ OD of duct.

Fig. 8-48

flange on square or rectangle *duct*, which is often made with heavy
black iron. The *frame* serves the same purpose as the *angle iron
companion flanges* used on *round pipe*. Matching holes are to be
drilled in these *frames* also, and a metal *pattern* should be made with
the *hole layout*, so that accuracy can be assured.

FOUNDATION MOUNTS or BOLT HOUSINGS

When a **foundation** or **pad** is required for the mounting of some
piece of *air conditioning* equipment, the *air conditioning contractor*
is usually required to both furnish the bolts and establish their
location for the *carpenter* who makes up the form.

Fig. 8-49 shows a cut-away view of one of the best *methods* of
preparing the bolts for this purpose. With the bolts being set in this
manner, it is possible to move any one of them by placing a nut on
the thread, then tapping with a hammer in any direction.

The shaded area represents the inside clear area around the bolt.
The size of the pipe stub, h, is not too important and it will be found
that a *pipe coupling* is excellent for this purpose, providing it does
not reach too far up into the thread as indicated at k. Most often, the
coupling can be cut in half, making up two mountings from the one
coupling.

238

When setting these mounts in the *form*, it is important that the tops of the *couplings* are kept flush with what will be the finished or top surface of the *pad*; also, that the thread of the bolt is slightly below the top of the *coupling* or *pipe* at *n*, with the length of thread extending above this point enough for the *frame* of the equipment, a flat washer, a lock washer and the required nut. Allow also a little extra thread for any variation that might be found in the *frame's* thickness.

The *weld* is made at the head of the bolt, *f*, under the plate, as shown. Any scrap plate is usable if the size approximates about 4″ x 4″. In fact, the use of scrap pieces of wide *angle iron* prove even better, as no side fins are necessary as they would be if using plate. The side fins, *j*, are welded to *coupling* or *pipe* as a precaution-

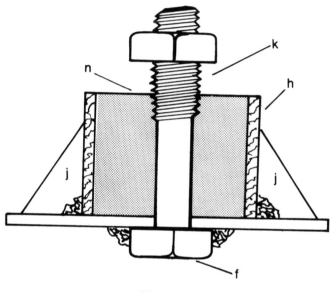

Fig. 8-49

ary measure, to prevent the bolt housing from turning in the *concrete* when tightening the nuts.

Many *shops* use only a piece of band or angle iron in a cross position on the shank of the bolt, which is welded into position. This is a lot faster to make up than the *method* such as given here. However, if the bolt should be out of line or shear off, there is no way to replace it. With the *method* shown here, if a bolt should shear off, it is possible to make a replacement with a new bolt welded in the same place.

Duct Layout and Fittings

TAKING RADIUS OFF SHEET

How can a **radius elbow cheek** be taken out of a sheet of metal, in one piece which would be too large if laid out in the usual manner?

This is possible only when the span across the proposed cheek, as indicated by line g in Fig. 8-50, does not exceed the width of the sheet. This can be determined, before attempting the *layout*, either by means of *trigonometry* or from a detail made to *scale*.

This method of layout is not recommended, due to the waste of material, and when possible should be developed in *gores*.

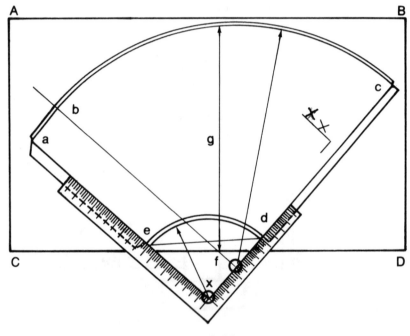

Fig. 8-50

Let A, B, C, D, in Fig. 8-50, represent the sheet of metal, with C-D as the bottom. Place the *square* in position as shown so that the radius required at e and d on the *square* sets directly above the bottom of the sheet.

Measure off the opening sizes a-e and c-d and move the *square* in either direction until the cheek is set on the sheet, using no more space than is necessary.

Mark points x and f and develop in the usual manner for the radius elbow.

chapter IX
ROOF JACKS AND
RELATED FITTINGS

FLAT ROOF
With Built-up Curb

This type of **roof jack** is made to fit a curb which is built around an opening through the roof, Fig. 9-1.

Although this curb can be made in various widths and heights, it is usually only a 2 x 4 set on edge, *c*, Fig. 9-1. This short height means that, since the roll of the roofing paper, *g*, must be allowed for, the apron cannot be more than 2-1/2″ overall, *f*, Fig. 9-1. The height of the 2 x 4, being 3-5/8″, does not leave too much clearance for the apron, even at 2-1/2″, having to consider that the roll of the paper will vary. For this reason, it is best to extend the width of the **jack,** so that the apron will be slightly beyond the roll of the paper.

In most cases, the *cutter* will not have to be concerned about this situation as the required dimensions will be given with a drawing on the *work sheet.*

However, there are occasions when the *cutter* is required to make a *roof jack*, being given only the actual size of the opening through the roof and the size of the riser duct which will project through the same opening.

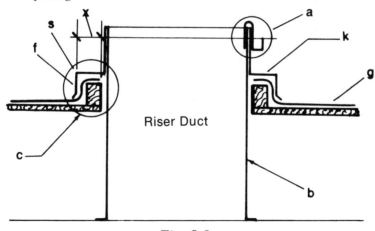

Riser Duct

Fig. 9-1

Roof Jacks

It is therefore necessary that the breaking point s, which is indicated in Figs. 9-1, 9-2 and 9-3, be made at a definite distance from the inside edge of the opening through the roof, as shown at x in Fig. 9-1.

Although the width of the 2 x 4 can easily be set at a definite figure, it is impossible to make a definite allowance for the roll or thickness of growth of the roofing paper.

The *author* has found it best to make a flat allowance of 3'' for x, Fig. 9-1, and by deducting the duct size from the opening size of the hole, it is a simple matter to make the overall measurement required for k.

Although this may be about an inch more on all sides than actually necessary, it is better to have the **jack** a little too large than too small.

The cut-away side view or elevation shown in Fig. 9-1 is for a flat roof. The *riser duct*, b, is brought through the roof and tied to the **jack** with a special 2'' government lock, which is shown at a.

The height above the roof will vary considerably, depending upon the designed layout on the roof. Usually having only a gooseneck set into the lock, the actual height is seldom too critical; but if an *elbow* with a continued run of duct is to carry on from the lock, a definite height is required in order to keep the run of duct at the proposed height off the roof.

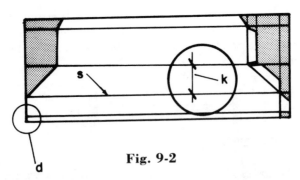

Fig. 9-2

The *layout* shown in Fig. 9-2 is a typical pattern for one side of a **jack**. If the width k is the same on all sides, this pattern will do for all four sides. Even if the actual size of the lock should change, as is the case were the duct rectangular, it would just be necessary to slide the *pattern* in or out to the desired size. The four required pieces can be blocked out at the shear, as shown in Fig. 9-2, the shaded area being the cut-away section, after the *layout* is completed.

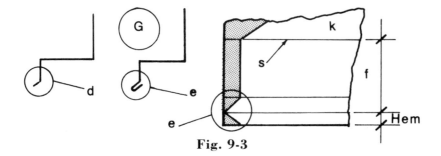

Fig. 9-3

When using the edge shown at *d*, the blocked-out piece will be as shown here. For the edge shown at *e*, in Fig. 9-3, 3/8″ more is required in the overall length of the blocked-out metal. The other edge is sufficient, due to the allowance made for the lap. The side view of *d* and *e* shown at *G*, clearly shows that there is no distinct advantage between either method, but rather a matter of choice by the *cutter*.

Notice, in Fig. 9-2, that the lap is left on only one end. The purpose of this is to enable the *cutter* to use the one pattern for all sides.

However, it is not always practical to have the lap on one end, especially on large *jacks* that are to be *spot-welded*, since the *jack* does not always fit into the *spot welding* machine. As a safety measure, the *cutter* can easily check this out before making up his *shear list*. Quite often the *fitting* is rectangular in shape and, by placing laps on the short side only, no difficulty will occur.

This same situation arises when making **canvas connections** and large **coil pans**.

Quite often, the *cutter* is required to make a multiple type **jack**, as shown in Fig. 9-4, where more than one riser pipe is carried up through the same opening. This is not really difficult to *lay out;* however, the fabrication problem must be given considerable thought.

No specific *method* of design can be given, as much depends on the *layout* of the holes and how the *fitting* can be made up at the seams.

If a spot-welded or riveted seam is desired, a lap must be left on the in-between sections, *q* and *r*, so that they can be made up from the outside. It is often possible to use the *pittsburg* lock, placing it also on the inner pieces, *q* and *r*, Fig. 9-4.

Roof Jacks

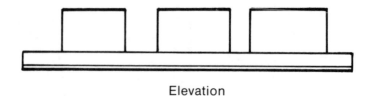

Elevation

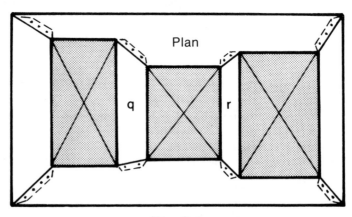

Plan

Fig. 9-4

PITCH ROOF
With Built-up Curb

The **jack** shown in Fig. 9-5 is the type used for a rounded truss roof or the roof with a given pitch, with both having a built-up curb.

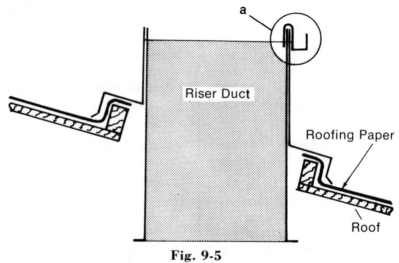

Fig. 9-5

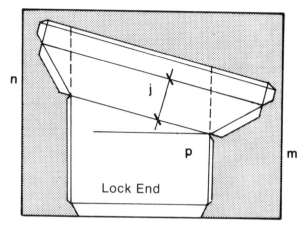

Fig. 9-6

The *layout*, in Fig. 9-6, is merely to show the *method* of developing and no specific dimensions, or pitch, are used.

The width of the *fitting* at the lock end is established on the sheet and a 1″ line, for lock allowance, is scribed across the bottom of the sheet. The throat, or short height, is placed as shown on the right side, indicated at *p*. A line is then carried horizontally across the sheet and off this line the pitch or rise is determined, which will give the line *n-m*, shown in Fig. 9-6.

With this established, the rest of the *layout* is handled in the same manner as described for Figs. 9-1 and 9-2. Likewise, the heel and throat will work up in the same manner; however, the forming will fit them to the pitch of the roof, as shown in Fig. 9-5.

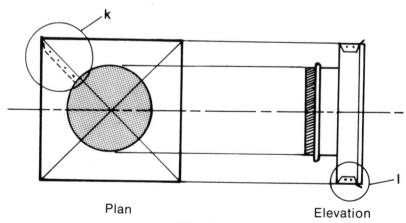

Plan   Elevation

Fig. 9-7

Roof Jacks

The type roof jack shown in Fig. 9-7 is usually made for a *rotary* vent or similar type. As a rule, they are not very large and can easily be made in one piece with practically no waste of metal.

By making up the *jack* in this manner, the metal can be blocked out at the shear and, since the height of the apron, *l*, is about 2″ or so, the edges can be scribed on all sides.

By crossing lines from corner to corner, the center is located. The desired size hole is marked and cut out, about 1/16″ over the required diameter.

The collar is made to required diameter and specified height and is usually dovetailed into place and soldered. Bead and crimped edge is also usually required as shown in the elevation, Fig. 9-7.

Referring again to the plan view, *k* shows how the *jack* is to be made in four pieces, if very large in size. It can be riveted or spot-welded; whichever way is chosen, the *jack* must be soldered watertight. When riveting, it is advisable to leave about 1/4″ lap on both corners of the miter and to place rivet holes directly on the miter line. This will assure accuracy and avoid the possibility of having the *fitting* buckle out of shape as sometimes happens when rivet holes do not line up.

SQUARE JACK
Using Pittsburg

A fast and simple *method* of making a square or rectangular **roof jack** is shown here. The *jack* is the same as shown on preceding pages, except that it is designed for roofs having openings without built-up curbs. Although many *cutters* prefer using the *spot welder* instead of this *method* (see next section), a neater job can be made by using the *pittsburg*. This will be particularly noticeable when soldering the *jack*.

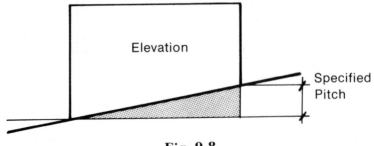

Fig. 9-8

Fig. 9-8 shows a side elevation of a roof *jack* as it would appear on the *work sheet*, showing the required pitch.

STEP 1. Assume the rectangle *A,B,C,D* is the metal to be used for the *layout*. It is usually best not to attempt to block out too close to size. It should be roughly figured and a piece of scrap metal can be found suitable for the purpose of developing this pattern.

As shown in Fig. 9-9, the *layout* is developed from the base of the sheet with point *a* located first. From point *a*, square up from the base of the sheet and measure off the given or required height for the throat, establishing point *b*.

STEP 2. At a right angle to line *a-b*, or parallel to it, draw a line longer than 12″, then intersect this same line with a 12″ mark measured from point *b*, to establish point *d*, Fig. 9-9.

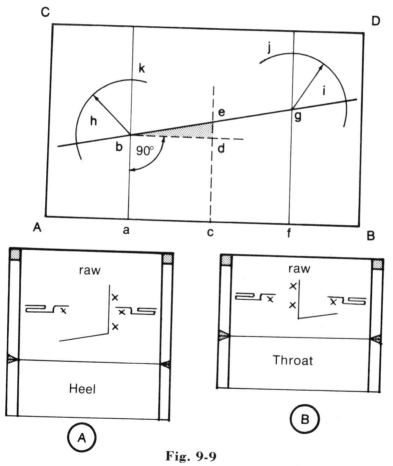

Fig. 9-9

Roof Jacks

STEP 3. Using a *square* off the base of the sheet from *c*, draw a short vertical line through point *d*. From point *d*, measure off the required amount of **rise** to locate point *e*, Fig. 9-9. Placing a *straightedge* at points *b* and *e*, draw a line diagonally across the sheet.

STEP 4. At the base of the sheet and from point *a*, measure off the required size of the opening on the cheek side to locate point *f*. Then, using a *square* off the base, draw a line up beyond the diagonal line almost to the top. This establishes point *g*.

STEP 5. Setting the dividers to the desired size of the apron or plate of the *jack*, scribe two arcs, using *b* and *g* as radius points, to establish points *h* and *i*, Fig. 9-9.

Placing a *straightedge* parallel to the diagonal line already on the pattern, and resting against the arcs as shown in Fig. 9-10, draw a line long enough to reach the outside of both arcs. This establishes points *k* and *j*. These points are placed on the edge for this purpose. In forming this *fitting*, it is necessary to mallet the 1/4'' edge over the leaf of the *brake*. These two points, *k* and *j*, are only as a convenience to the *brakeman*, making it easier to line up the edge.

STEP 6. From points *h* and *i* and at a right angle to that line, draw connecting lines to square off the apron and determine points *m* and *n*.

Place the 1/4'' lines at the heel and throat and, at *h-b* and *g-i*, allow a 5/16'' edge.

As these two edges lie flat in the *pittsburg*, it is better to have them 1/16'' wider, making it easier to hold in place.

STEP 7. The throat and heel are made up in the usual manner, as shown at *A* and *B*. Notch as shown, with the *pittsburgs* being cut back 1'' at the raw edge.

To complete the cheek, cut out and mark for the usual breaks. Use this as a pattern and cut out another cheek, marking on the opposite side since a right and left are required.

SQUARE JACK
Lap and Spot Weld

The actual development of this **jack** is the same as described for the previous *method*, using the *pittsburg*, shown in Fig. 9-9. A few

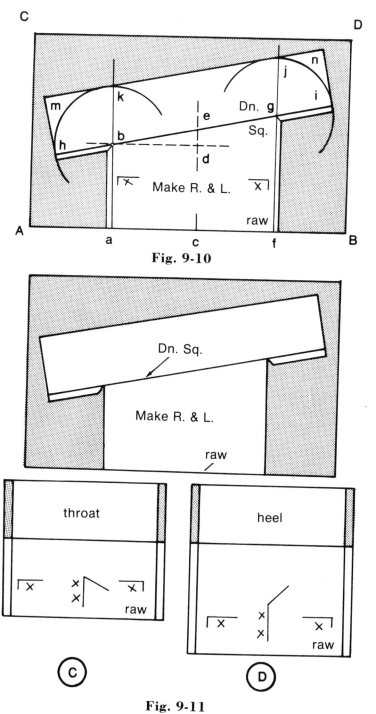

Fig. 9-10

Fig. 9-11

Roof Jacks

changes are easily seen by comparing the two *layouts*, Figs. 9-10 and 9-11, the latter being for the spot-welded *jack*.

For the cheeks, the heel and throat are left raw. The edges on the apron, *h-b* and *g-i* of Fig. 9-9, can be made about 1/2″ instead of 5/16″ as allowed here.

For the heel and throat (wrappers), the edge of the apron is left raw with the lap on sides made 5/8″, *C* and *D*, instead of 1″ as required for the *pittsburg*.

The laps should, whenever possible, be made as described, having the cheeks placed into the *spot welder* horizontally. However, the size of the *fitting* will be the determining factor, that is, whether or not the *fitting* can be held under the arm of the machine.

If necessary, the laps should be reversed. It is important that the *cutter* check this point before making his *layout*.

The *cutter* will find that the amount of rise or pitch line will be given in one of the following ways:

Rise per foot.   (2′-0″).
Rise per width of the *fitting*.
Rise per specified degree.
Rise per given measurement, heel and throat.

JACK
With Flat Cap

The actual *layout* of this **jack** is similar to the **Square Roof Jack** to a given pitch, which is shown elsewhere in this book (see **Index**). A slight difference is indicated at point *m*, Fig. 9-13, where the top

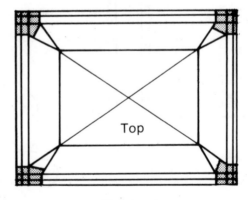

Fig. 9-12

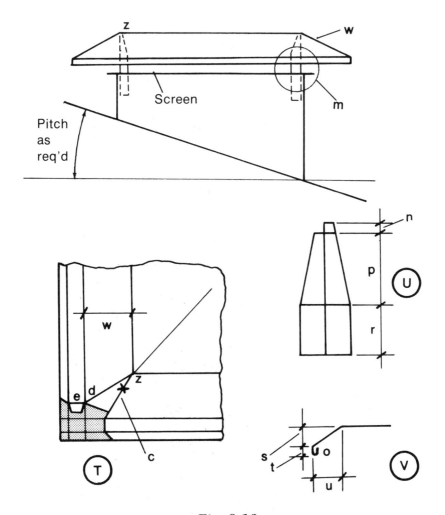

Fig. 9-13

edge is turned out 1/2" at 90°, which acts both as a stiffener and a place to tack-solder the screen.

The **cap** *layout* is made in one piece, as shown in Fig. 9-12, with the flat inside area being the same size as the opening of the **jack.** To develop the outer edge *w*, make the *layout* a 2" in 4" pitch and, as shown at *V*, *u* is 4", *t* is 1", *s* is 2", *o* is 1/2". Proportioning in this manner, the slant height of *w* is a scant 4-1/2"; however, 4-1/2" is satisfactory for this *layout*.

Roof Jacks

Shown at *T*, which is a *blow-up* corner of the **cap** *layout*, is the *method* of notching for the corners. By cutting back 1/2'' each way and making connecting lines to the corners, from points *c* and *d*, the miter lines are developed. Allow a small tab, *e*, to tie the corners together. The shaded area is notched out as shown, with a straight cut made to the corner point *z*.

U shows the corner leg *layout* of 1'' x 1'' angle, made with about 22 gauge metal. The tab, *n*, at the top is 1/2'' and, extending through the cap, locks it in place. Make the height *p* 4-1/2'', which is usually the required allowance to keep the opening between the **cap** and **jack** of correct proportions in relation to the allowed overhang. Height *r*, at *U*, is 2'', with the angles being spot-welded into place in each corner.

After the legs and screen are in place on the *jack*, set the **top** on the legs and mark each corner for the required slot. Using a screwdriver to make the slots, mount the **top** and solder each tab. This same type of **jack** can be made with the legs cut out directly from the body of the **jack.** However, this method is rather impractical, since it requires a considerable amount of extra time.

Shown at *W*, in Fig. 9-14, is a **cap** which is slightly different, but made in somewhat the same manner, and also used upon the same

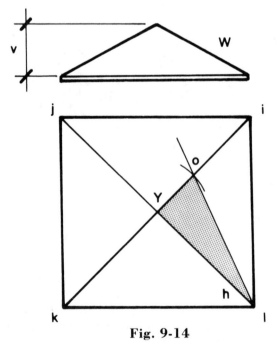

Fig. 9-14

type of **jack.** The height, v, represents the same height as set off in the plan view, Fig. 9-14, at point o. The length of the line o-l is then used as the radius, from point Y of Fig. 9-15, to establish the circle shown.

Set *trammel points* equal to the length of one side in Fig. 9-14 and step off the various points as shown, h, i, j and k. Connect all points with straight lines, as shown, also to the corner point Y. Add the necessary allowance that is required for the seam, apron, and hem, for a complete pattern. The shaded area represents the metal roughed out for the *layout*, c,d,e and f. This is, of course, the excess which must be cut from the pattern.

When very large or rectangular **caps** of this same type are required, they are made in separate sections, with standing seams. The *layout* then becomes similar to that applied when making a *hood* or *canopy*.

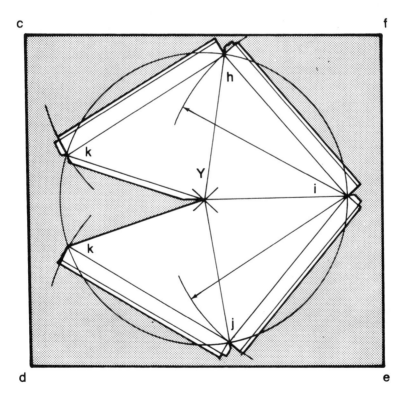

Fig. 9-15

Roof Jacks

JACK
Round, to a Given Pitch

STEP 1. Lay out the roof line and section *A*, as shown in Fig. 9-16, establishing points *1, 5, 7, 11*.

STEP 2. Scribe the half-circles *B* and *C*, dividing them into an equal number of parts, Fig. 9-17, and numbering each piece as shown, *1-5* and *7-11*.

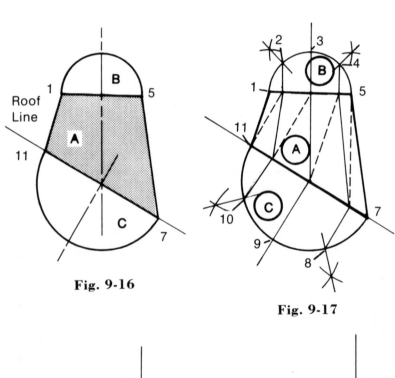

Fig. 9-16

Fig. 9-17

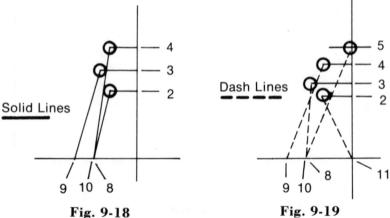

Fig. 9-18

Fig. 9-19

STEP 3. Carry these points directly to lines *1-5* and *7-11*, then join these points with connecting lines. To make it easier when making the *layout*, it is best to make each group of lines different, such as a solid for the lines working up from numbers *8-11* and broken lines for those working down from numbers *2-5*. This is shown in Fig. 9-20.

STEP 4. The true length of these lines must now be determined, in the usual manner, through the process of *triangulation*. As shown in Figs. 9-18 and 9-19, the development of each line is at a set distance off the vertical line or, as it should rightly be called, the **center** line of the *fitting*. The horizontal lines are therefore placed, in this manner, off the center in order to correctly determine the difference between the top and bottom openings.

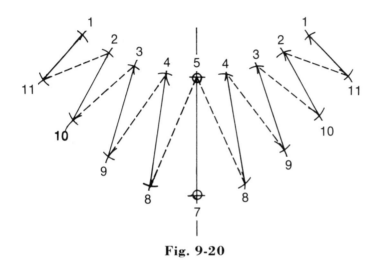

Fig. 9-20

This is easier to understand if the *reader* will visualize the top opening as the size of the pipe being carried directly through the *fitting*. The difference between this pipe and the bottom of the *jack* is the actual growth, of course, variable at each point around the circumference of the flared base, shown in Fig. 9-17.

STEP 5. Starting with line *5-7*, which is the true length, placed at any convenient place on the sheet, proceed to lay out the *pattern* in the usual manner, Fig. 9-20. Take the span between *7-8*, etc., from the half-circle *C*, and *4-5* from the half-circle *B*, Fig. 9-17.

STEP 6. Allow the necessary material for the required edges. Cut a square piece of metal for the plate, allowing 4″ to 6″ beyond the diameter of the opening on the roof or pitch line.

Roof Jacks

NOTE: Although this type of layout is very accurate, considerable time is required in the actual layout. For this reason, it is more practical to use the **rollout method** *shown below.*

Most *shops* have patterns on hand of various sizes and pitch. From these *patterns*, one can be found that is close to the size required. With a few changes, a satisfactory *pattern* can usually be made.

STEP 7. Having established the points shown in Fig. 9-20, the faired lines are drawn in the usual manner. Also, add about 5/8″ to the *miter line* for attaching to the *roof plates.*

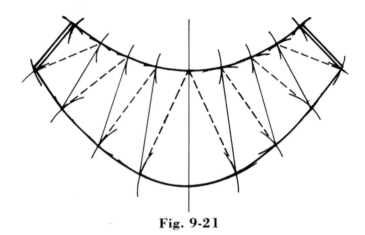

Fig. 9-21

Allow, at each end, for the seam to complete the *pattern layout,* Fig. 9-21.

NOTE: For a more completely detailed description of the triangulation method as applied here, see **Index** *under* **Triangulation.**

JACK
Round, Rollout Method

STEP 1. Lay out the elevation of the *roof jack* as shown in Fig. 9-22, with the line *d-c* drawn to the required pitch. Also, scribe the half-circles and cut out the entire *pattern.*

STEP 2. Break up 90° on lines *a-b* and *c-d*, thus forming the half-section.

Check the squareness of the half-circles and, if necessary, place

256

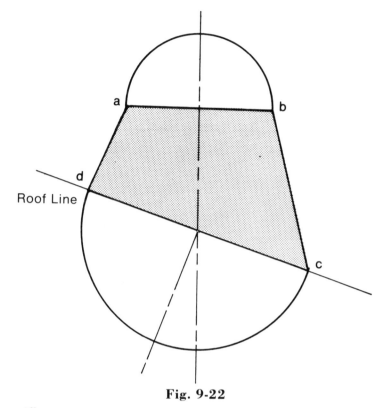

Fig. 9-22

a stiffener or gusset where required to keep square. This section is now ready to roll, with the procedure the same as described in the Wye Branch Rollout Method (see **Index**).

FILTER BOX, SMALL PLENUM
Outdoor, Watertight

Many of the smaller *air conditioning installations* are built up from a *unit* containing a *blower* and *coil*, to which a *filter section* is added on the suction side of the *blower*.

There are unlimited designs in which the *filters* are installed in these small *plenums*. Each is just a little bit different from the others, enough to make it impossible to use one standard design that would be suitable for each situation.

This added *filter section* usually has a limited space in which it can be made, such as in a *line* of *duct* itself, with the space available seldom, if ever, being large enough to accommodate the required *filters*. However, when the *filters* are to be used in a small prepared *plenum*, separate from the actual *duct line*, the situation

Roof Fittings

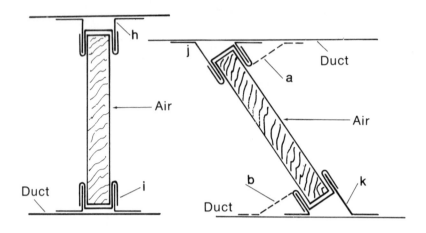

Fig. 9-23   Fig. 9-24

is generally just the opposite: there is too much space and it requires blocking off the area around the *filters*.

The latter is undoubtedly the easier problem to overcome. However, regardless of what the problem may be, it is up to the *cutter* to use his own ability in handling the situation. Although no definite *layout* can be prescribed, there are certain basic *methods* of framing the *filters* and either problem can be easily worked out.

As shown in Fig. 9-23, a *filter* is used in a line of *duct* with ample space for vertical framing, as indicated at *h* and *i*. As a comparison, Fig. 9-24 shows the same size *filter* as installed in a smaller *duct* and attention is called to the variation in the framing, at *j* and *k*. Also note that the dotted portion, shown at *b* and *a*, is an alternate way of making the frames.

Although these *frames*, or *channels* as some refer to them, can be made up in one piece of metal, it is not practical due to the difficulty encountered in forming. It will also be easier to change the various pieces around, should any error show up in regards to length or height of any of the *frames*. The design shown in Fig. 9-27 is easily used for V-type layouts, such as shown in Fig. 9-26. They can easily be installed with *pop rivets*, as indicated at *e*, *f* and *g*, Fig. 9-27. Referring to Fig. 9-26, when designing the *layout* for this a type of *filter bank*, it is a practical idea to use two strips of metal, *c*, cut out to the size of the *filters* and, by placing them in various positions, one will save considerable time in developing the best means of placing the *filters*.

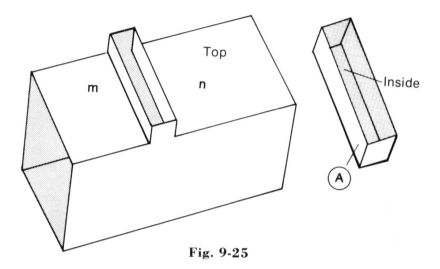

Fig. 9-25

The *filter box* shown in Fig. 9-25 is one of the best designs for outdoor use, where a weatherproof design is required. With this *layout*, no *rain* can possibly enter the box and be carried into the *duct line*. The sides of the box are attached to the bottom and formed in a

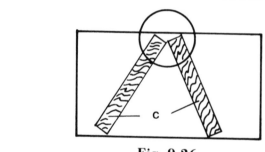

Fig. 9-26

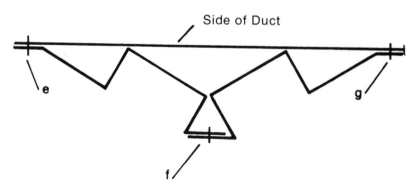

Fig. 9-27

three-sided section. The top pieces, *m* and *n*, have the *pittsburg* on each side, so that these are the same as the throats of a square *elbow*. The *cap* shown at *A* is made large enough to slip on the top opening and, as it fits down almost to the top of the box, no rain can possibly enter inside. The necessary *framing* is required inside to enclose the *filter*; however, none is required across the top as the raised opening is used for this purpose. It may require extension of the bottom frame to raise the *filter* up enough to reach inside this opening. When the plenum is used outdoors, it is best to have the opening on top, as it is easier to keep water-tight than when a side opening is used. In reference to side openings, whenever two *filters* are used together in line, it is usually necessary to place a pull strap on the bottom *channel* with the far end of the strap being turned up in order to withdraw the *filters* which must be changed periodically. When making the opening for the *filter*, turn a 1/2″ of metal to the inside to act as a stiffener for the opening.

The usual *throw-away* type of *filter* is of a rather flimsy construction and has a tendency to bulge through the middle. Be sure to allow enough width at the opening slot so that these flimsy filters will not catch hold of the edges when being removed.

WEATHERPROOF COVER

This item, shown in Fig. 9-28, is a **protective covering,** against

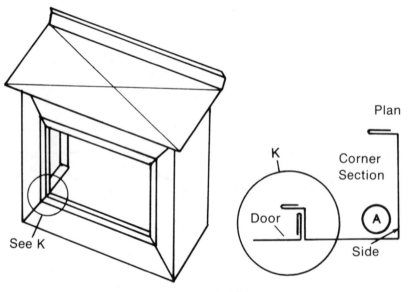

Fig. 9-28

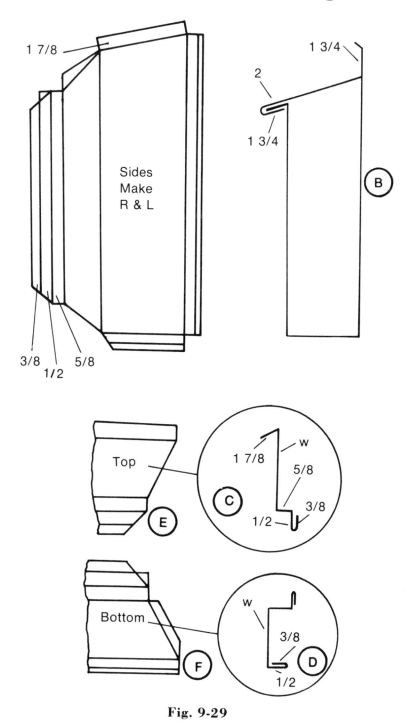

Fig. 9-29

Roof Fittings

the *weather*, for *meter boxes, gauges*, etc. which must be mounted on an outside wall. The *layout* is not difficult and requires no back or bottom. As the overall dimensions will vary, no figures besides those for the *door* framing are shown. Either *sash locks* or *hinges* can be used on the *door* which is the regular pan type (see *blow-up A*).

The diagrams shown are self explanatory, Fig. 9-29, being a **side** pattern which requires two pieces, both a **right** and **left.** Make the top pitch about 1″ or 2″, as the depth is usually only 8″. The *plan view* of *A* shows the forming process of the **sides,** with detail of *door* and *frame*. Note that the flange with hem is turned to the inside for mounting the unit.

The *side view*, in *B*, shows the flange for the top made with a slight feathered edge to serve as a groove for *mastic*. The profile of the front across the top is shown at *C*, with a corner of the *pattern* in *E*. These *frame* figures are, of course, the same for all four pieces; however, *w* is the unknown for both the top and bottom. In the same manner, *D* and *F* are shown for the bottom.

The top is heavily panelled with a 2″ overhang on the three sides, shown at *B*. The *door* is 9/16″ deep, unless rubber stripping is to be used around the frame; if so, allow 1/2″ for depth.

STORM COLLAR

Several variations can be made on the type of **storm collar** shown here without affecting its purpose. This is one of the *routine fittings* which must be given as little time in developing as possible. For this reason, most *shops* keep a *pattern* for the particular *layout* they prefer on their *jobs*.

A close similarity will be found in the *layout* of a **round roof cap** and a **storm collar.**

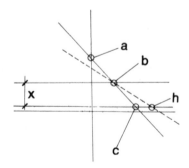

Fig. 9-30

On the other types of *collars*, the changes would be as follows: The *pitch line*, *a-c* Fig. 9-30, may be changed or the *collar* lengthened out (dashed line). Also, the standing seam may be omitted and spot-welded instead. The *collar* shown here has a pair of screws to draw it up tight when in place, as shown in Fig. 9-32. The *crimped* edge can also be omitted, as well as the small edge turned at the top.

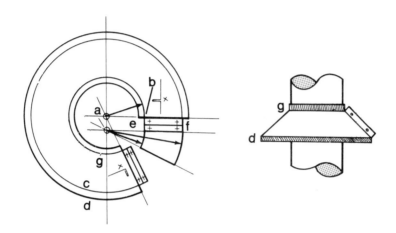

Fig. 9-31   **Fig. 9-32**

STEP 1. Lay out the *elevation* to the given measurements, as shown in Fig. 9-30 at *b* and *c*; this establishes radius point or *apex a*.

STEP 2. Using *a* as center and *b*, *c* as radius, scribe the circles shown in Fig. 9-31.

STEP 3. Refer to the **Circumference Chart,** using the diameter at *b*, Fig. 9-30, and from this deduct about 1/2" (depending on how much draw-up is desired). From a set point on the circle, *b* Fig. 9-31, measure off this amount. Drawing a line through both points will determine the cut-out.

STEP 4. Add the necessary standing seam and allowance for edges *d* and *g*, Fig. 9-32. Where the space is too small to lay out the standing seam, as shown at *e-f*, Fig. 9-31, cut off the edge at *e-f* and make this half of the standing seam separate. After forming, this can be spot-welded to the *collar*.

The height *x*, shown in Fig. 9-30, can be made shorter. However, the taper should be spread out as shown by the line *b-h*, which changes the *apex a*.

Roof Fittings

STANDARD GLOBE VENT

The following pages show a simplified *method* of proportioning for a **standard globe vent.** An accompanying Chart, Fig. 9-36, gives the necessary dimensions for twelve of the sizes most often used.

By using the figures given in this Chart, it is only necessary to add the material required for seams, hems, etc., requiring no actual *layout* by the *cutter*. It will, of course, be necessary to lay out the bolt holes. The *cap* will also require a bit of *layout*, but this is not much to consider, since it is very little work.

The *method* of proportioning is the most important part of a **globe vent** *layout* since the actual *fitting* is only composed of three major pieces, namely: the **stem, cap** and **wind guard.** As the *stem* is a given diameter at the start of the *layout*, this leaves only the *cap* and *wind guard* to be developed.

Since these pieces should have a definite proportion in relation to the *stem* diameter, an accurate *method* must be used to determine each piece.

Shown here, in Figs. 9-33 through 9-37, is the *method* used in developing the Chart, with a step-by-step description of the same. Following this progressive *layout*, the *cutter* can determine the required sizes for developing any size of **globe vent.**

As a means of quickly and accurately following the procedure described, each point is lettered in progressive order as it is located and placed on the drawing.

For explanatory purposes, this *layout* is made full size for a **globe vent** having a 3″ diameter *stem*. The development shown is a half-profile, which is all that is ever necessary for this purpose.

The *method* used here in the following procedure is proportioned in accordance to the diameter of the *stem*.

By observing the left side of the Chart, it will be noticed that there are three fractions shown, with the specific purpose indicated that each one is to serve. These fractions, marked *A, B, C,* must be calculated, as an error here would throw the entire *layout* out of proportion.

This fact will become more apparent to the *reader* as he works up the following step-by-step procedure.

STEP 1. Draw horizontal line *X-Y* at random length, Fig. 9-33. Also erect a vertical line which will represent the center line of the *stem* and locate the intersecting point *a*.

STEP 2. From point *a*, measure off, to the right, one-half the width of the *stem*, thereby locating point *b*. At this point *b*, draw a vertical line through this point, extending downward to about the desired length of the *stem*.

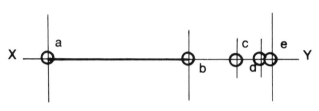

Fig. 9-33

STEP 3. From point *b* on the line *X-Y*, measure off to the right to establish the points *c*, *d*, and *e*, Fig. 9-33. These points represent the three fractions taken from the Chart. The point *c* is 1/16 of the *stem* diameter, which is *B*, the overhang of the hood.

Point *d* is 1/4 of the *stem* diameter, which is *A*, the **rise** of the *hood*. Point *e* is 1/8 of the *stem* diameter, which is *C*, the space between the edge of the *hood* and the *wind guard*, Fig. 9-33.

Erect a vertical line through point *e*, at a random length, for use later in the development.

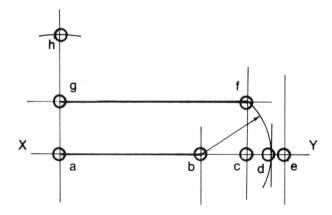

Fig. 9-34

Roof Fittings

DIAMETER OF STEM	RISE OF HOOD	OVERHANG OF HOOD	SPACE BETWEEN EDGE OF HOOD AND WIND-GUARD	STEM		
				CIRCUMFERENCE	HEIGHT	AREA IN SQUARE INCHES
	A-1/4	B-1/6	C-1/8			
6	1-1/2	1	3/4	18-7/8		28-1/4
8	2	1-3/8	1	25-1/8		50-1/4
10	2-1/2	1-11/16	1-1/4	31-7/16		78-1/2
12	3	2	1-1/2	37-11/16		113-1/16
14	3-1/2	2-3/8	1-3/4	44	ANY LENGTH CAN BE USED	153-15/16
15	3-3/4	2-1/2	1-7/8	47-1/8		176-11/16
16	4	2-11/16	2	50-5/16		201-1/16
18	4-1/2	3	2-1/4	56-9/16		254-1/2
20	5	3-3/8	2-1/2	62-13/16		314
24	6	4	3	75-7/16		452-7/16
30	7-1/2	5	3-3/4	94-1/4		706-7/8
36	9	6	4-1/2	113-1/16		1017-7/8

Fig. 9-36

STANDARD GLOBE VENT

HOOD		WIND-BAND		
RADIUS	CIRCUMFERENCE	CIRCUMFERENCE	DIAMETER	WIDTH
4-5/16	25-1/8	29-7/8	9-1/2	2-1/4
5-3/4	33-13/16	40-1/16	12-3/4	2-13/16
7-1/8	42	49-7/8	15-7/8	3-3/4
8-9/16	50-1/4	59-11/16	19	4-7/16
10-1/16	58-15/16	69-15/16	22-1/4	5-1/8
10-11/16	62-7/8	75	23-7/8	5-3/4
11-7/16	67-3/16	80-1/8	25-1/2	5-15/16
12-3/16	75-7/16	89-9/16	28-1/2	6-3/4
14-3/8	84-1/16	99-3/4	31-7/16	7-3/16
17-1/8	100-9/16	119-7/16	38	8-15/16
21-5/16	125-5/8	149-1/4	47-1/2	10-13/16
25-3/4	150-3/4	179	57	12-1/2

Fig. 9-36

Roof Fittings

STEP 4. With the *dividers* set to the width of *b-d*, scribe an arc as shown in Fig. 9-34. Erect a vertical line from point *c* upward through the arc to establish point *f*, which is actually the corner or outer edge of the *cap*. Note that the point *d* is placed on line *X-Y* for convenience only since the arc is all that is required.

From the point *f*, place a horizontal line that intersects the center line to establish point *g*. This line, *f-g*, is one-half of the *cap* and the line *a-b*, directly below, is one-half of the *stem*, Fig. 9-34.

With the *dividers* still set to width *b-d* and using point *g* as radius point, scribe a short arc to cross the center line for point *h*, as shown in Fig. 9-34.

STEP 5. As shown in Fig. 9-35, draw a diagonal line from point *h* through point *f* to locate point *j*. This will be the center of the *wind guard*. Draw a diagonal up from point *b* through point *f* to intersect the line carried up from *e*. This will establish the point *i*, or the top of the *wind guard*.

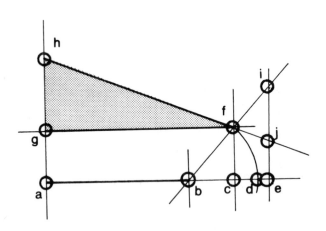

Fig. 9-35

STEP 6. Using *j* as center or radius point, set the *dividers* to the width of *i-j*, and swing arc to locate point *k*, which is the bottom of the *wind guard*, as shown in Fig. 9-37.

From points *i* and *k*, carry over a horizontal line to intersect the vertical center line thereby establishing points *l* and *m*.

STEP 7. On the center line, from points *a* and *b*, measure down-

268

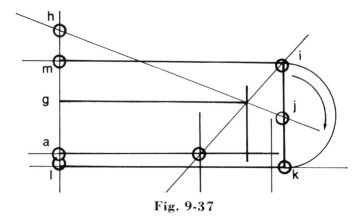

Fig. 9-37

ward for the desired length of the *stem*, which will locate points *n* and *o*, Fig. 9-38.

Connecting lines from *a* to *o*, *b* to *n*, and *n* to *o* will complete one-half of the *stem*.

This also completes the proportioning *layout* and the *cutter* can prepare his *shear list*. In taking the various measurements off the *layout* just completed, also the dimensions from the Chart, it must clearly be understood that no allowances have been made for any seams or hems.

It is usually best to rivet the seam on the *wind guard*, as this guard is run through the machine to put 2 or 3 beads around it for the

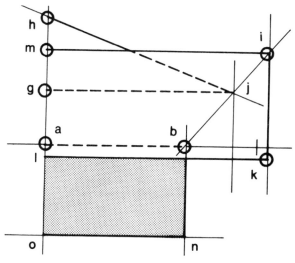

Fig. 9-38

Roof Fittings

purpose of stiffening. The riveted seam, therefore, makes it easier to handle. The *rivet* holes can be laid out directly on the line, only having to allow about 3/8''; even 1/4'' will do on certain sizes. This is to be added to both ends; therefore, it is 1/2'' or 3/4'' to the circumference. It is also best to hem both top and bottom, making it necessary to add about an inch to the width, as well.

The *cutter* must use his own ability in making the allowances for hems and seams, since the different size *fittings* and gauge material used will be a deciding factor.

The *cutter* must make the *hole layout** for the *wind guard* in the flat. The larger *fittings* will require four gussets or brackets, while the smaller ones need only three.

This brings forth the question of using either gussets or brackets. With this question arises the need for an added *storm collar, q,* to the same *fitting,* as shown enlarged in Fig. 9-40 at *A.*

As is the case with many of the various items the *cutter* makes, there are always one or two other ways in which the item can be made.

The *author* prefers the added *storm collar* and then uses the band-iron brackets, such as shown in Fig. 9-39 and Fig. 9-40 at *C.* The use of gussets rules out the *storm collar,* although it is not impossible to be made with it.

The gusset design is shown on the right side of the vent appearing in Fig. 9-39, while bracket is shown on the left side for comparison.

When the added *storm collar* is desired, it can readily be made by

*Holes for bolts that fasten to gussets for brackets.

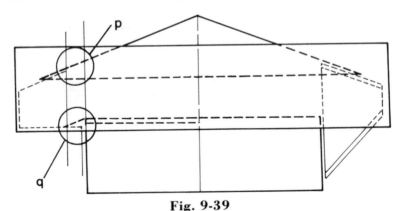

Fig. 9-39

270

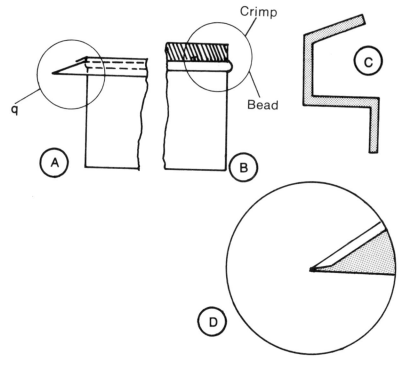

Crimp

Bead

Fig. 9-40

taking the dimensions directly off the *cap*, such as shown at *p* in Fig. 9-39. In order to fit on *stem*, the *collar* must be made a trifle larger; also, the *stem* should have the added edge as shown at *A*, Fig. 9-40.

The edge must be hemmed if no *storm collar* is to be used; also, if possible, put a bead around the *stem*, Fig. 9-40 *B*.

The *stem* should be seamed by running it through the **acme** machine. In making the *cap*, the usual *method* is applied, the seam being either riveted or possibly spot-welded. In either case, the seam should be skim-soldered underneath. This radius is the one indicated on the chart, with the *circumference* being the outside *stretch-out*, as usual. See Fig. 9-40, *D*.

GOOSENECK DETAIL

Although the actual *layout* of a **gooseneck** is comparatively simple, care should be taken in the construction of this type of *fitting*.

Roof Fittings

As the *gooseneck* is exposed to the weather, it is important that every precaution be taken to prevent the possibility of rain entering inside the *fitting*, where it could travel down the *riser pipe*. Since the *gooseneck* is often visible from the street, consideration should be given to its design and fabrication.

Theoretically, the correctly designed *gooseneck* is shown in Fig. 9-42, with the throat and heel radius being carried around to 135°, making it almost impossible for rain to ever blow into the *riser duct*. When finished, this also makes the best looking *gooseneck*.

However, this radius type of *fitting* does involve more work than the square design shown in Fig. 9-43. This is especially so when the *fitting* is very large. The radius *fitting* costs considerably more to make than the square type.

Both *fittings* are very simple to make, since they are regular 90° elbows with an added 45°, to make 135° overall, as shown in Figs. 9-42 and 9-43. However, the *layout* does become extremely awkward when very large in size. Although the cheeks of the smaller *fittings* are easily made in one piece, it is not practical on the larger *goose-*

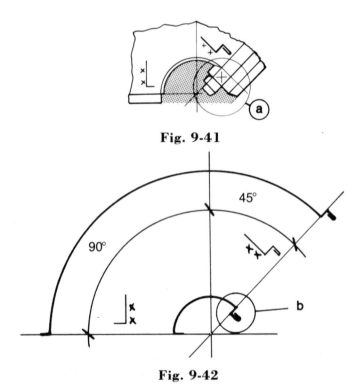

Fig. 9-41

Fig. 9-42

neck, even when possible. It is definitely a better practice to make the cheeks in two *gores.* This gives added strength to the *fitting.*

When making the cheek in *gores,* the standing seams should be run vertically or diagonally, avoiding horizontal seams if possible. When the horizontal or diagonal seam must be used, be certain to keep the double part of the standing seam on top. This applies also to a seam that is placed on the side of the heel wrapper, such as shown in Fig. 9-43 at *f.*

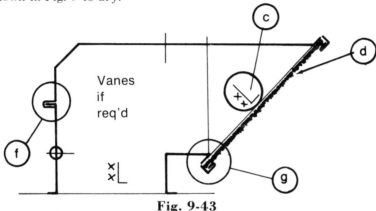

Vanes if req'd

Fig. 9-43

In order to make a tight seal, notch the corners, as shown in Fig. 9-44 where the method of notching is shown at *e.* The diagonal dotted line merely shows the corner of the double part of the seam as it should be left in the flat. When the seam is put together, this tab is folded around to make a tight and neatly finished corner.

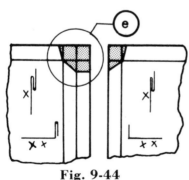

Fig. 9-44

There is a choice as to the *method* of applying the wire mesh or hardware cloth on the end. In some cases, the wire is clamped directly into the *fitting.* For this *method,* a cup-lock with a hem is left on the actual pieces of the *fitting* when the *layout* is being made, Fig. 9-41 at *a.* The forming of the edge is best shown at *b* in Fig. 9-42.

Roof Fittings

Again referring to *a* in Fig. 9-41, notice that the metal is left extending over 1″ at the throat. This same amount must be left on the heel also. The heel and throat wrappers, having the *pittsburg*, are not required to have this added 1″ piece, thereby making it possible to use the *pittsburg* lock machine. When made up in this manner and the corners are soldered, the mesh can be clamped into the frame, making a very neat finish.

This *method* works very well on the smaller *fittings*. However, it is rather impractical on the larger *goosenecks*, with a separate frame being more satisfactory.

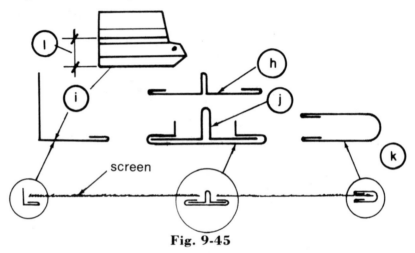

screen

Fig. 9-45

By making the screen in a separate frame, the *cutter* has a choice of two methods of constructing the frame. One method is shown at *k* in Fig. 9-45, whereby a U-type frame (1″ wide) encloses the wire mesh. This can be placed in the cup-lock, as shown at *b* in Fig. 9-42, and held in place with screws.

The second method fits over a 7/8″ flange that is left on the *gooseneck*, as shown at *c* in Fig. 9-43. This type of frame is easily made from lock stock, which is shown at *i* in Fig. 9-45, by trimming to 2-1/4″ from the outer edge, as shown in *l*.

The frame is easily riveted together and the wire mesh tack-soldered into place. Holding the frame into place over the flange is done with either metal screws or bolts, *g*.

For very large screens, a center support is required. This support can be made in several ways. Two very simple means are shown in *h* and *j*. These methods are practical from the standpoint that they not only make a good stiffener, but also provide a solid edge to tack solder the screen, as shown in Fig. 9-45. These cross bars should be used vertically on the *fitting*, using two bars for very large openings.

The mesh is usually purchased in 36″ and 48″ widths, which will therefore limit the *cutter* to its use, without splicing.

CHIMNEY HOOD OR CAP

Although this type **cap** or **hood** is not used as extensively as the newly designed, more efficient, approved types available, there are a considerable number still being put into use.

Due to the simplicity in *layout* and speed with which it is fabricated, it is well to know the actual *layout* process.

STEP 1. Calculate the blockout size *R, S, B, D,* having the length *B-D* equal to the *circumference* of the *pipe,* plus allowance for seam.

The height *R-B* is 2/3 of the *circumference* plus only 1/2″ for the top seam. Scribe across top and sides for seam allowance.

STEP 2. Lay off piece of pipe *A-C,* into 6 equal parts, with line *A-C* being 1 division, or 1/6, of the *circumference* off the bottom. The division points then are marked *A, 1, 2, M, 3, 4,* and *C.*

STEP 3. With the *dividers* set from *M* to *A* and using *M* as center, scribe the arc *A* to *N.*

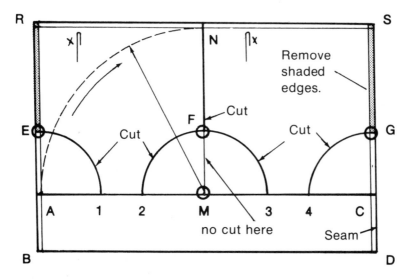

Fig. 9-46

Roof Fittings

STEP 4. Then scribe arcs *E-1*, *F-2*, *F-3* and *G-4*, with their respective centers as shown.

STEP 5. Draw the straight lines *E-R*, *N-F* and *S-G*, which completes the *layout*.

STEP 6. Cut on the lines marked *Cut* in Fig. 9-46, adding the necessary laps as shown, for seam. The *fitting* is now ready to form.

*NOTE: The top seam should **not** be formed until after the fitting has been rolled and the side seam is fastened. The top seam is then formed, with one-half rolled each way, thus making the half-round top, as required.*

TRANSITE HANGERS

This **clamp-type bracket** is another of the so-called *little items* which the *cutter* is required to make. While it appears to be most insignificant, the *cutter* will find that it requires a lot of handling and to do a good job in a hurry is a real challenge as well as a real work-out, when having several to make.

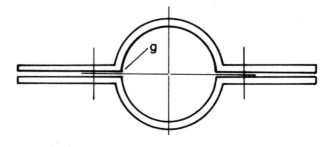

Fig. 9-47

This is best laid out by drawing a circle on a piece of scrap metal, the OD size of the *transite*. Then, by making an allowance of no less than 1/2″ between the arms, such as shown at *g*, in Fig. 9-47, the *bandiron* can be fitted quite accurately. As this type of hanger is usually made with light 1/8″ stock, it is easily rolled and, by having the break marks previously laid out, the arms can then be shaped as shown. Most *shops* have an old *brake* that can be used for this purpose, or it can be done in a *vise* with a little more effort and a heavy *hammer* or *single jack*.

Be careful to allow sufficient gap so that a tight clamp will be made when the bolts are drawn up. The required length of the arms is usually given, unless the opening through the *slab* is only slightly larger, in which case any length will be sufficient, n, Fig. 9-48.

One bolt on each side is usually used, about 3/8'' diameter; how-ever, whenever the brackets are requested with long arms, it will take two.

The same brackets are often used for the *transite* rising on the outside of the building, R in Fig. 9-48, in which case hanger straps will be needed to reach the side of the building.

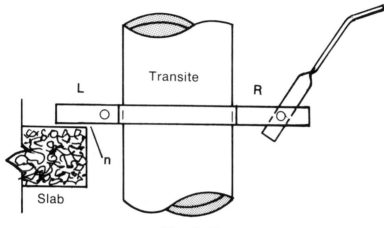

Transite

L

R

n

Slab

Fig. 9-48

In most cases, these strap hangers will be a given length and also be required to have a full or half-twist between each end, so that they will lay flat against the building and also fit the brackets.

How the hangers lay on the *slab* is shown at L of Fig. 9-48, with R showing the hanger in position. On small *transite*, use a wrap-around (16 or 18 gauge) hanger which only needs one bolt, having only one opening.

BUCKET DETAIL

This type of *fitting* is another of the many items the *cutter* is often required to make. Although referred to as a **bucket,** it is actually a sleeve that is placed through the ceiling, usually being the connect-ing point between the galvanized *vent pipe* and *transite* that ex-tends upward inside the *can* or pipe part of the *fitting*.

Either the *square* or *round* type will do the job; however, as each *shop* has its own *method* of how it should be done, the *cutter* seldom has to make any decision.

Roof Fittings

The necessary dimensions are given, but it is usually left to the *cutter* to make the *layout* of the holes in the *plate* which fits against the *ceiling*.

The *square* type, shown in Fig. 9-49, is made up a lot faster than the *round* type in Fig. 9-50. This is, of course, due to the extra work in fabricating, as the actual *layout* time is about the same.

To make the *bucket* shown in Fig. 9-49, make a stub of pipe the given length and opening size, making one end raw with the other end having a 7/8'' flange turned out.

The end plate should be blocked out 3-3/4'' above the pipe size. This will make the flange 1'' wide, with 7/8'' to be hemmed over the edge, or flange, of the pipe.

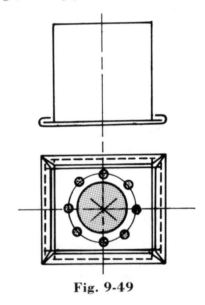

Fig. 9-49

Scribe a 7/8'' edge around all four sides and, using diagonal lines from corner to corner, locate the center and scribe a clearance hole for the required metal pipe. Make a check as to the actual thickness of the *transite*, then lay out the series of small holes, as shown.

As a safety measure, to assure that the holes are clear of the *transite*, keep the small holes as far to the outer edge as the pipe above will allow. These holes can be 3/8'' diameter unless otherwise specified.

The fabrication is very simple, with a final clamping of flange edge made in the *brake*.

The hole spacing usually works out a few inches apart, as shown in Figs. 9-49 and 9-50.

The procedure is almost the same for the development of the *round* type shown in Fig. 9-50, with a few necessary exceptions. As shown in the *blow-up, A*, it will be noticed that the edge is turned out about 3/16″ on the stub. It is usually best to mark out the plate after the *can*, or stub, has the 3/16″ edge turned out, by placing it on the plate and scribing all around the outer edge. Then add the additional 3/16″ stock to turn up in the **easy edger,** as shown in *a*, Fig. 9-50.

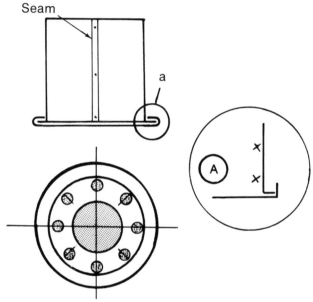

Seam

a

A

Fig. 9-50

The *reader* must remember that this *fitting* is one of the routine *fittings* which must be put through as quickly as possible. For this reason, all work must be kept to a minimum.

Lay out holes in the same manner described for the *square bucket*. Do not cut the large opening until after the *fitting* is fabricated as, in most cases, there is very little stock left on the plate, making it difficult to turn the edge.

When fabricating the *bucket*, turn the 3/16″ edge up square, place the *round* pipe in, then peen the edge over slightly all the way around. The edge can then be squeezed tight in the *brake*, after which the *fitting* is completed by cutting the hole.

Roof Fittings

LOUVER LAYOUT

Although **louvers** are, in most cases, made to a specified size and frame design, a basic *method* can be applied in their development.

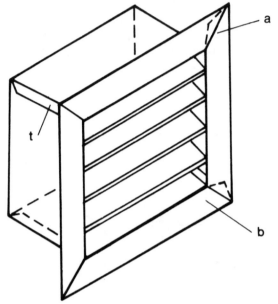

Fig. 9-51

Fig. 9-52 is a cut-away side *elevation* of a *louver*, such as shown in Fig. 9-51. The *layout* shows the *blades* as being equally spaced to the top of the *louver*; however, this will seldom be the case. The *blades* are spaced up equally from the bottom, with the height of the *louver* determining the top *blade*. For a good example of this, see Fig. 9-54.

The *method* of *blade* spacing and *layout* shown here is the preferred *method* of the *author*, as it provides the maximum protection from rain.

Although the *blade layout* offers the maximum in protection, maximum free or open area is not obtained. However, most *louvers* are made well over the actual size required, unless a space problem is encountered. In many cases, where a *louver* is protected by buildings or a *parapet wall*, considerably more freedom is possible in the *louver's* design. As shown in Fig. 9-53, the dash line *w* indicates manner of *blade layout* where the top of the lower *blade* is in line with the *blade* above. This same *layout* is applied in Fig. 9-54.

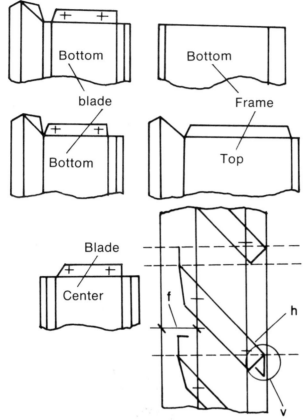

Fig. 9-52

Shown in Fig. 9-53 is the *layout* as it appears when a variation is made in the *blade layout*. Here the width of the flange is left between the *blades*. This permits a slightly greater free or open area between each *blade*, which would still be laid out in the same manner, using a *square* as shown in Fig. 9-54.

Also shown at *v* is how the edges can turn up parallel with the *blades*, instead of being squared off as shown in Fig. 9-53 at *z*.

The *layout* in Fig. 9-54 represents the metal as blocked out for the *sides*, which requires two pieces, with the *layout* as follows:

STEP 1. Establish the line *s-t* equal to the width of the *blades*. As shown here, the *blades* are finished flush with the front. When set in any amount, the *s-t* line must also move in accordingly.

STEP 2. Using the *combination square* off the edge of the sheet, draw in the *blades*, *r*, Fig. 9-54.

Roof Fittings

STEP 3. The rivet lines are drawn in parallel to the *blades* at 3/8″ from these *blades lines*. Be certain to turn the bottom flanges upward as a convenience when riveting and soldering.

This is the **only** portion of the *louver* where the laps are turned up instead of down. Laps *a*, *b*, *q*, *t* and *y* are all turned down.

When the *blades* are to be spot-welded in the *louver*, no rivet lines are required; but, for accurate placement of the *blades*, it is most practical to prick mark the corner points of each *blade* line, Fig. 9-54.

If of light metal, the center *blades* can be made 1/8″ less in width, including the top *blade* which will vary in its stretchout. In a few cases, it will be possible to use a center *blade* with a slight revision similar to the profile shown at *S*.

Actual allowance for growth of *blade* lengths will depend on the gauge of metal and on how sharp a break is made on the ends. The latter is a case of making a test break on the *brake* that will be used.

The edges which form the top and bottom of *blades* are usually made 1/2″ and 3/8″, except the two places shown at *d* and *e*, Fig. 9-53.

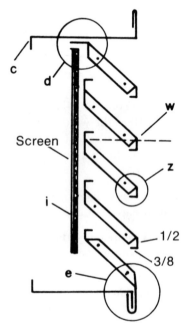

Fig. 9-53

When facilities make it possible, the louver can be spot-welded together. However, where heavy metal has been used, or the *louver* is very large, it should be riveted together. Care must be taken when laying out the rivet holes, particularly in the *blades*.

As indicated at *h* in Fig. 9-52, unless the space marked *f* is moved proportionately inward, the holes in the *louver blades* must be of two different measurements. When laid out in this manner, the *blades* must all be formed with the same side up, since the holes will not be interchangeable. This can readily be done; however, it does leave considerable chance for error when being formed.

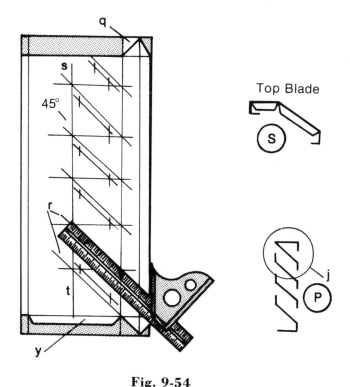

Fig. 9-54

Practically all *louvers* require wire mesh or hardware cloth on the back, as indicated at *i* of Fig. 9-53. Usually 1/2'' mesh is used; it is, however, often required that 1/4'' mesh be used.

This mesh should be tack-soldered on, making certain the acid is washed off when finished as rust will form rapidly. Many *shops*, that take pride in the quality of work they turn out, will have the soldered spots touched up with an *aluminum* paint spray bomb.

Roof Fittings

When a small, narrow *louver* is required, the *layout* and *fabrication* can be cut to a minimum. As shown at *P*, both the top and bottom *blades* can easily be made up in one piece as part of the frame. For added area, turn in a 3/8″ edge and omit the usual 1/4″ edge normally turned to the inside.

In cases where the *louver* is extremely long, it should be made up in two sections. When the *louver* is large, but not actually large enough to be made in sections, a center partition or division can be built through the center. This requires extra care when allowing for the growth of metal, as well as with the *laying* out of the rivet holes.

Actual requirements are usually specified on the *order sheet*, along with the gauge metal required.

For *louvers* that are built with long *blades*, it is best to place a center brace vertically on the center and front of the *blades*.

For this brace, use either galvanized band-iron or regular hanger strap. Tie this brace to the *louver* with a metal screw through the bottom edge of each blade. The top *blade* can often be tack-soldered to the frame, as shown at *d*, Fig. 9-53. By turning about 1/2″ out on the wire mesh, this can also be tack-soldered in place at the same time. This makes it unnecessary to form any break on the *blade*.

METAL DUCT COVERING

The insulation of *duct* on the outside, which is exposed to the weather, requires some form of protection. One *method* that is used quite often is described here, where an outer layer of light metal is formed up to cover the insulation.

As shown in Fig. 9-55, the top and both sides are wrapped to form a water-tight covering.

Although the covering shown here is pictured as being closely *fitted*, this is not usually the case. The measurements given the *cutter* are seldom accurate, particularly on old *duct-work*. Therefore, some clearance should be allowed.

The panels are made with standing seams that are extended across the top to both of the outer edges, as shown at *A*, with the *layout* for notching the corners being shown at *j* and *k* in Fig. 9-56. For this *method* of notching and with the seams being out to the edges, a water-tight seal is made. Shown at *a*, in Fig. 9-55, is a *2 x 4* which can be placed through the center of the *duct line* to keep the metal covering from sagging and forming hollows which will hold water.

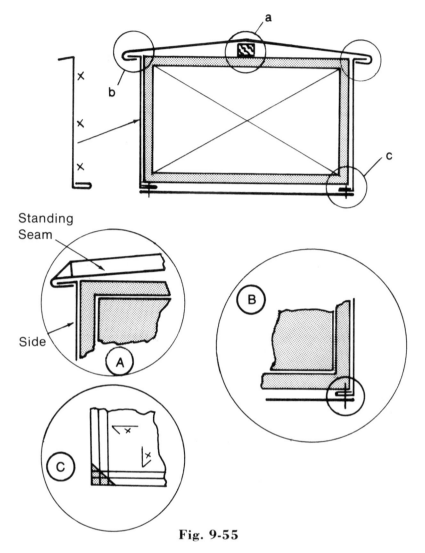

Fig. 9-55

Fig. 9-55 also shows how the side panels are formed. The panels should be made as long as possible. As shown at *b*, in Fig. 9-55, the overhang of the top is 1″ and the corners being finished, as indicated at *C*. As shown at *f*, in Fig. 9-57, diagonal seams are good and, for any seams other than vertical, keep the double edge of the standing seam down from the top of the *duct, g.*

About the only difficult places to properly close are the vertical corners, such as the heels and throats as shown at *h*. For these particular places, the regular *pittsburg* lock is used, placing it on the throat and heel.

Roof Fittings

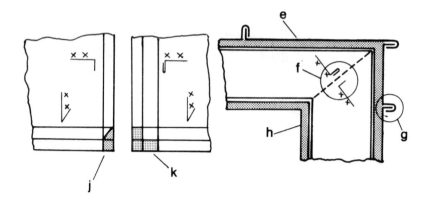

Fig. 9-56   **Fig. 9-57**

Referring again to Fig. 9-55, the corner at the bottom, as shown at *c*, is a hemmed edge, turned 1-1/4″ underneath the *duct*. Using either a hanger strap or prepared metal angles, these are screwed to the bottom edge, *c*, being spaced a few feet apart, such as shown at *B*.

The various pieces should be carefully marked, with the *layout* being indicated on a section of the covering or the *work sheet*, showing the location of each piece. Most of the *duct* covering jobs take a considerably large amount of metal and, when it is formed up, the pieces all look very much alike and require considerable time to sort out.

On this type of work, it is a good practice to start at a given point and refer to the various pieces as either being for on *top* or the **right** or **left side,** as required. Use an identifying number or letter in successive order, such as *L-1, R-1, L-2, R-2, T-1*, etc.

The pieces made for the sides should all be heavily panelled or cross-broken. However, panelling the top pieces causes a pocket or hollow to form on each side of the standing seams, which naturally fills up with water.

The use of the center brace does help considerably in overcoming this. Although there still is some difficulty encountered, it is not enough to cause any problem.

chapter X

TRIGONOMETRY

TRIGONOMETRY

No attempt will be made here to explain **trigonometry** in its entirety as this, in itself, would fill the book. However, for the benefit of the *reader* who may not understand this branch of *mathematics*, a resume of the subject is given here.

The use of **trigonometry** has been mentioned throughout this book as being a definite advantage when applied to certain problems encountered in *layout* work. The *methods* required in the development of these problems will be shown here.

This branch of *mathematics* is used to determine the unknown parts of a *triangle*. It is only necessary to know the **sides** of any *triangle* to calculate the **angles,** or knowing the **angles** to calculate the **sides.**

In Fig. 10-1 the **angles** are indicated by the capital letters A, B, C and the **sides** of the triangle are indicated by the small letters a, b, c.

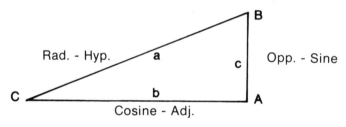

Fig. 10-1

The calculations deal with the **ratios** between these **sides,** which are expressed more often as the **functions** of the **angles,** with each *function* being named in reference to the *ratio* it represents.

Another fact regarding the **right triangle** is that, no matter how the length of the sides may vary, no change occurs in the *ratios* of the *sides* **if** the acute *angle* remains unchanged. Also, if two *sides* are known, the remaining *side* and the *angles* may be found; or if only one *side* and the *angles* are known, the remaining *sides* may be determined.

Trigonometry

The **ratio** or **functions** of the *angle* can be determined through the various *sides*, which are related to *angle C* as follows:

a=hypotenuse   radius C
b=side adjacent   cosine C
c=side opposite   sine C

These various **functions** of **angles** are developed by means of the following **formulas:**

sine	$=$	$\dfrac{c}{a}$	or	$\dfrac{\text{side opposite}}{\text{hypotenuse}}$	or	$\dfrac{\text{opp.}}{\text{hyp.}}$
cosine	$=$	$\dfrac{b}{a}$	or	$\dfrac{\text{side adjacent}}{\text{hypotenuse}}$	or	$\dfrac{\text{adj.}}{\text{hyp.}}$
tangent	$=$	$\dfrac{c}{b}$	or	$\dfrac{\text{side opposite}}{\text{side adjacent}}$	or	$\dfrac{\text{opp.}}{\text{adj.}}$
cotangent	$=$	$\dfrac{b}{c}$	or	$\dfrac{\text{side adjacent}}{\text{side opposite}}$	or	$\dfrac{\text{adj.}}{\text{opp.}}$
secant	$=$	$\dfrac{a}{b}$	or	$\dfrac{\text{hypotenuse}}{\text{side adjacent}}$	or	$\dfrac{\text{hyp.}}{\text{adj.}}$
cosecant	$=$	$\dfrac{a}{c}$	or	$\dfrac{\text{hypotenuse}}{\text{side opposite}}$	or	$\dfrac{\text{hyp.}}{\text{opp.}}$

These *formulas* will determine the *functions* of *angle C* and, because any *function* of an acute *angle* also equals the *co-function* of the **complement** of that *angle*, the *formulas* used for developing *functions* of *angle C* are also used for developing the *functions* of *angle B* as follows:

sine C = cosine B
cosine C = sine B
tangent C = cotangent B
cotangent C = tangent B
secant C = cosecant B
cosecant C = secant B

Because the **ratio** or **function** is constant, it is possible to use the **Table of Functions** is developing various problems without having to make calculations to determine the value of each *function* (see elsewhere in this Chapter).

In applying the *formulas* for the *sides*, the various *sides* are either multiplied or divided by a *function value*, or *factor*, taken from the

Table of Functions in accordance to the **degree of angle** being developed, as follows:

To find side a, the
length of the hypotenuse:

Multiply the side opposite by the *cosecant* or . . .
Multiply the side adjacent by the *secant* or . . .
Divide the side opposite by the *sine* or . . .
Divide the side adjacent by the *cosine.*

To find side b, the
length of the side adjacent:

Multiply the side opposite by the *cotangent* or . . .
Multiply the hypotenuse by the *cosine* or . . .
Divide the side opposite by the *tangent* or . . .
Divide the hypotenuse by the *secant.*

To find side c, the
length of the side opposite:

Multiply the hypotenuse by the *sine* or . . .
Multiply the side adjacent by the *tangent* or . . .
Divide the hypotenuse by the *cosecant* or . . .
Divide the side adjacent by the *cotangent.*

The foregoing *formulas* are for the development of the **sides,** if one of the *sides* and an *angle* are known.

The problem shown in Fig. 10-1, which is to be 22-1/2°, is shown again in the shaded area of Fig. 10-2 so that the *reader* can compare this *fitting* with the various *functions* when placed in their relative positions. For this particular problem, Fig. 10-3, we only require the development of the *miter line.*

A very simple means of doing this is to apply the *tangent function.* Select the *factor* for a 22-1/2° *angle*, from the Table of Functions, a **value** of .4142, and multiply this by the radius of the *fitting*, which is also the *hypotenuse, a.* This is clearly shown in Fig. 10-2. The radius for this particular problem is 40″ and, multiplied by .4142, results in a product of 16.568 or 16-9/16 inches. By measuring up 16-9/16″, at 40″ from point k, a connecting line through these determined points develops the **miter line.**

It is not particularly necessary to select any specific figure for the radius since it only requires the correct *function value* for the *degree of angle.* When multiplied by any figure, the product would be

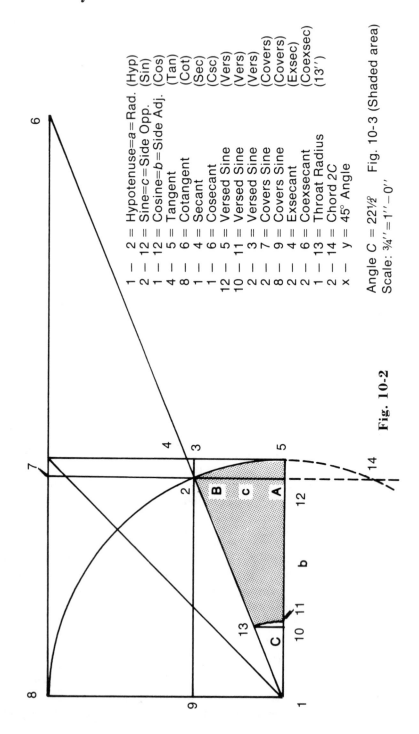

1 — 2 = Hypotenuse=a=Rad. (Hyp)
2 — 12 = Sine=c=Side Opp. (Sin)
1 — 12 = Cosine=b=Side Adj. (Cos)
4 — 5 = Tangent (Tan)
8 — 6 = Cotangent (Cot)
1 — 4 = Secant (Sec)
1 — 6 = Cosecant (Csc)
12 — 5 = Versed Sine (Vers)
10 — 11 = Versed Sine (Vers)
2 — 3 = Versed Sine (Vers)
2 — 7 = Covers Sine (Covers)
8 — 9 = Covers Sine (Covers)
2 — 4 = Exsecant (Exsec)
2 — 6 = Coexsecant (Coexsec)
1 — 13 = Throat Radius (13″)
2 — 14 = Chord 2C
x — y = 45° Angle

Angle C = 22½°   Fig. 10-3 (Shaded area)
Scale: ¾″=1″−0″

Fig. 10-2

on the same *miter line* when measured up at the distance chosen from the *apex* or *miter line*, which is out from point k, Fig. 10–3.

Knowing the *hypotenuse, radius,* and *side a* are one and the same, it is obvious that any *fitting* having a required radius has this *side* figure already given with no calculating being required. This also is true for the *angles C* and *B*.

The required *degree* of *angle*, which is angle *C*, is shown on the *work sheet* with the problem. Knowing angle *C*, we also know the angle *B*, its *complement*.

Each *fitting* is easily developed from these two known facts. As mentioned earlier in this section, if only one of the **sides** and an **angle** are known, the remaining *sides* may be determined.

When the various *function values* are determined, it is then possible to calculate the length of each line that is associated with each *function* by multiplying each *factor* by the *hypotenuse* or *radius*, which is 40'' in the example *fitting* shown in Fig. 10-3. These calculations can be checked through the scaled detailed show in Fig. 10-2.

Sin	$\dfrac{c}{a}$	*.3827* x 40 = 15.308 or 15-5/16''
Cos	$\dfrac{b}{a}$	*.9239* x 40 = 36.956 or 37''
Tan	$\dfrac{c}{b}$	*.4142* x 40 = 16.568 or 16-9/16''
Cot	$\dfrac{b}{c}$	*2.414* x 40 = 96.56 or 96-9/16''
Sec	$\dfrac{a}{b}$	*1.082* x 40 = 43.28 or 43-1/4''
Csc	$\dfrac{a}{c}$	*2.613* x 40 = 104.52 or 104-1/2''
Vers	$\dfrac{a-b}{a}$	*.0761* x 40 = 3.044 or 3''
Covers	$\dfrac{a-c}{a}$	*.618* x 40 = 24.72 or 24-11/16''
Exsec	$\dfrac{a-b}{b}$	*.0811* x 40 = 3.244 or 3-1/4''
Coexsec	$\dfrac{a-c}{c}$	*1.61* x 40 = 64.4 or 64-7/16''

NOTE: The italicized **decimal** *is the calculated function value, which can be taken either from a Function Chart or derived*

291

Trigonometry

TRIGONOMETRIC FUNCTIONS

Angle	Sin	Cos	Tan	Cot	Sec	Csc	•
0°	.0000	1.0000	.0000	∞	1.0000	∞	90°
1°	.0175	.9998	.0175	57.2900	1.0002	57.2987	89
2	.0349	.9994	.0349	28.6363	1.0006	28.6537	88
3	.0523	.9986	.0524	19.0811	1.0014	19.1073	87
4	.0698	.9976	.0699	14.3007	1.0024	14.3356	86
5°	.0872	.9962	.0875	11.4301	1.0038	11.4737	85°
6	.1045	.9945	.1051	9.5144	1.0055	9.5668	84
7	.1219	.9925	.1228	8.1443	1.0075	8.2055	83
8	.1392	.9903	.1405	7.1154	1.0098	7.1853	82
9	.1564	.9877	.1584	6.3138	1.0125	6.3925	81
10°	.1736	.9848	.1763	5.6713	1.0154	5.7588	80°
11	.1908	.9816	.1944	5.1446	1.0187	5.2408	79
12	.2079	.9781	.2126	4.7046	1.0223	4.8097	78
13	.2250	.9744	.2309	4.3315	1.0263	4.4454	77
14	.2419	.9703	.2493	4.0108	1.0306	4.1336	76
15°	.2588	.9659	.2679	3.7321	1.0353	3.8637	75°
16	.2756	.9613	.2867	3.4874	1.0403	3.6280	74
17	.2924	.9563	.3057	3.2709	1.0457	3.4203	73
18	.3090	.9511	.3249	3.0777	1.0515	3.2361	72
19	.3256	.9455	.3443	2.9042	1.0576	3.0716	71
20°	.3420	.9397	.3640	2.7475	1.0642	2.9238	70°
21	.3584	.9336	.3839	2.6051	1.0711	2.7904	69
22	.3746	.9272	.4040	2.4751	1.0785	2.6695	68
23	.3907	.9205	.4245	2.3559	1.0864	2.5593	67
24	.4067	.9135	.4452	2.2460	1.0946	2.4586	66
25°	.4226	.9063	.4663	2.1445	1.1034	2.3662	65°
26	.4384	.8988	.4877	2.0503	1.1126	2.2812	64
27	.4540	.8910	.5095	1.9626	1.1223	2.2027	63
28	.4695	.8829	.5317	1.8807	1.1326	2.1301	62
29	.4848	.8746	.5543	1.8040	1.1434	2.0627	61
30°	.5000	.8660	.5774	1.7321	1.1547	2.0000	60°
31	.5150	.8572	.6009	1.6643	1.1666	1.9416	59
32	.5299	.8480	.6249	1.6003	1.1792	1.8871	58
33	.5446	.8387	.6494	1.5399	1.1924	1.8361	57
34	.5592	.8290	.6745	1.4826	1.2062	1.7883	56
35°	.5736	.8192	.7002	1.4281	1.2208	1.7434	55°
36	.5878	.8090	.7265	1.3764	1.2361	1.7013	54
37	.6018	.7986	.7536	1.3270	1.2521	1.6616	53
38	.6157	.7880	.7813	1.2799	1.2690	1.6243	52
39	.6293	.7771	.8098	1.2349	1.2868	1.5890	51
40°	.6428	.7660	.8391	1.1918	1.3054	1.5557	50°
41	.6561	.7547	.8693	1.1504	1.3250	1.5243	49
42	.6691	.7431	.9004	1.1106	1.3456	1.4945	48
43	.6820	.7314	.9325	1.0724	1.3673	1.4663	47
44	.6947	.7193	.9657	1.0355	1.3902	1.4396	46
45°	.7071	.7071	1.0000	1.0000	1.4142	1.4142	45°
•	Cos	Sin	Cot	Tan	Csc	Sec	Angle

through the formula which is shown preceding the figure.

The following is the development of the **sides** according to the *formulas* given earlier in this section, and applied to the same sample problem in Fig. 10-3. Here also the *functional values* are italicized.

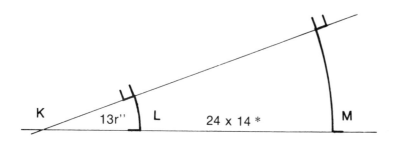

Fig. 10-3

Length of Hypotenuse   *Side a or radius*   **C=22-1/2°**
2.613 x 15.3 = 39.97 or 40''   side opp. x csc
15.3 ÷ *.3827* = 39.9 or 40''   side opp. ÷ sin
1.082 x 37 = 40''   side adj. x sec
37 ÷ *.9239* = 40''   side adj. ÷ csc

Length of Side Adj.   *Side b or cosine*   **C=22-1/2°**
.9239 x 40 = 36.95 or 37''   hyp. x csc
40 ÷ *1.082* = 36.9 or 37''   hyp. ÷ sec
2.414 x 15.3 = 36.93 or 37''   side opp. x cot
15.3 ÷ *.4142* = 36.9 or 37''   side opp. ÷ tan

Length of Side Opp.   *Side c or sine*   **C=22-1/2°**
.3827 x 40 = 15.3   hyp. x sin
40 ÷ *2.613* = 15.3   hyp. ÷ csc
.4142 x 37 = 15.3   side adj. x tan
37 ÷ *2.414* = 15.3   side adj. ÷ cot

APPLYING THE VERSED SINE *VERS*

The cheeks of the various angles are usually cut from scrap metal at the bench. However, there are occasions where the *cutter* may block out the material at the shear along with his regular *shear list.* Although this needs only a roughed-out piece of metal a little larger

Trigonometry

than necessary, it does require a calculated measurement.

To find the block-out size of material required for a cheek of the sample *fitting* in Fig. 10-3, we must determine the extreme points of the cheek. This can be best shown by referring to the shaded area of Fig. 10-2, which is the cheek of our sample problem. The extreme outside points are *13* to *5*, and *2* to *5*, not including any edge allowance.

It is obvious that the *sine 2-12* is the height, with the *vers sine 10-11* plus size of the cheek points *11-5*, being equal to the width.

The *versed sine* is found by using the same *function value* here that was used to develop the span of *12-5*. However, instead of multiplying by 40'', use the throat radius which is 13''. The result of this calculation is .9893, or 1''. This, added to the cheek size of 27'', makes the width 28''. The height will be found by multiplying the radius, 40'', by the *factor* or *function value* for this *degree* of *angle*, or .3827. The result is 15.308, or 15-5/16''. By adding an inch or two, which is according to the flanges and edges required, we will have the block-out size sufficient for the cheek *layout*.

This same process can apply to any similar type of *fitting*. Shown here are several of the most frequently used *degrees of angles*, in *chart* form, together with the necessary *sine* and *versed sine factors*. As these calculations do not require an accurate figure, the decimal figure of the *sine function* can be interpreted as a slightly larger fraction, such as shown here.

The *cutter* must remember that an allowance must be added to these calculations for necessary *flanges*, etc. As this amount will not have to be exact, it is best to allow a little more than the determined figure.

DEGREE	VERSED SINE	SINE	APPROX.
15°	.03407	.2588	5/16''
20°	.06031	.3420	3/8''
22-1/2°	.07612	.3827	7/16''
30°	.13397	.5000	9/16''
45°	.29289	.7071	3/4
60°	.5000	.8660	7/8''

In determining the height of the large *gore* in the *layout* of a *gored radius elbow* (see **Index**), the *sine function* must be used here also. However, where the openings of the *elbow* are considerably different in size, this calculation alone will not give an accurate height. Although this measurement is only to determine the width of the metal

294

required, it is often necessary to make a more accurate calculation, which may involve far more figuring. This can be eliminated by making a *scaled detail* of the *fitting*, thereby saving considerable time.

A similar problem is found in *Taking Radius Off the Sheet* (see **Index**).

Here, also the *vers function* is required at the throat, using the *factor* of the 45° angle. Apply the *versed sine* in the same manner at the throat, with the span across the center of the cheek being taken from the given sizes of the *fitting*. However, when any extreme difference is found between the openings, it will then require additional calculating to obtain this measurement. Although it can be figured that the span across the center will be about 1/3 less than the difference of the openings, it is merely a rough estimate. Here, again, is a place where a *scaled detail* will save time.

Another example of applying the *versed sine* is shown in Math Problem No. 16, for the purpose of determining the length of **straight duct** required between two *angles* to develop an *offset* (see **Index**).

The following **formulas** have been grouped together for the purpose of convenience.

RIGHT TRIANGLE

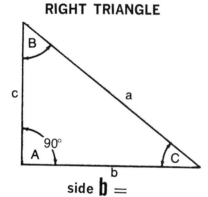

side $a =$

b × secant C	$\dfrac{b}{\text{cosine C}}$
b × cosecant B	$\dfrac{b}{\text{sine B}}$
c × cosecant C	$\dfrac{c}{\text{sine C}}$
c × secant B	$\dfrac{c}{\text{cosine B}}$

side $b =$

a × cosecant C	$\dfrac{a}{\text{secant C}}$
a × sine B	$\dfrac{a}{\text{cosecant B}}$
c × cotangent C	$\dfrac{c}{\text{tangent C}}$
c × tangent B	$\dfrac{c}{\text{cotangent B}}$

side $c =$

a × sine C	$\dfrac{a}{\text{cosecant C}}$
a × cosine B	$\dfrac{a}{\text{secant B}}$
b × tangent C	$\dfrac{b}{\text{cotangent C}}$
b × cotangent B	$\dfrac{b}{\text{tangent B}}$

Trigonometry

$$\frac{c}{a} = \text{sine C} \qquad = \text{cosine B}$$

$$\frac{b}{a} = \text{cosine C} \qquad = \text{sine B}$$

$$\frac{c}{b} = \text{tangent C} \qquad = \text{cotangent B}$$

$$\frac{b}{c} = \text{cotangent C} = \text{tangent B}$$

$$\frac{a}{b} = \text{secant C} \qquad = \text{cosecant B}$$

$$\frac{a}{c} = \text{cosecant C} \qquad = \text{secant B}$$

C angle **B**

Although the various *formulas* are shown for the **oblique triangle,** no discussion on applying them has been given here.

It is possible to solve most *oblique triangles* by dividing them into two right triangles. This can usually be done by drawing a perpendicular line through the center. However, while this is possible, it is usually impractical to apply in actual *shop* practice, since this involves considerable calculating.

OBLIQUE TRIANGLES

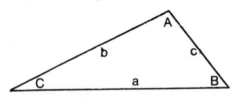

FIND	GIVEN	FORMULA
b	a,B,A	$\dfrac{a \times \sin B}{\sin A}$
c	a,A,C	$\dfrac{a \times \sin C}{\sin A}$
c	b,C,B	$b \times \sin C \times \csc B$
A	B,C	$180° - (B + C)$
B	A,C	$180° - (A + C)$
C	A,B	$180° - (A + B)$

FIND	GIVEN	FORMULA
tan A	a,C,b	$\dfrac{a \times \sin C}{b - (a \times \cos C)}$
cos A	a,b,c	$\dfrac{b^2 + c^2 - a^2}{2bc}$
sin B	b,A,a	$\dfrac{b \times \sin A}{a}$
cot B	a,C,b	$\dfrac{a \times \csc C}{b} - \cot C$
sin C	c,A,a	$\dfrac{c \times \sin A}{a}$

chapter XI

MATH PROBLEMS

INTRODUCTION

The following **math problems** are those most frequently used by the *cutter* in the *layout of air conditioning fittings*. These problems have been simplified as much as possible so that they can be readily applied by the *cutter* in actual *shop* practice, even though his knowledge of *mathematics* may be limited.

In order to apply *mathematics* effectively, it is necessary that we not only know **how** to develop the various *methods*, but that we also know **where** they are best applied. Unless we are able to use the most direct solution to any specific problems, more time is lost than gained.

Calculations for any problem should be kept to a minimum whenever possible, i. e., the *cutter* should avoid making a lengthy process of developing figures which have little or no direct bearing on the actual problem.

The *cutter* will find it most helpful to purchase both a Mathematics Manual and a *plumber's rule* which has a 45° *scale*. This *scale* determines the *hypotenuse* lengths of 45° *angles* having legs of equal length. A few *trigonometry* problems have been placed in the Math Section, rather than in the Chapter on that subject. The purpose in doing this is to have these particular problems where they may be handier to find, being shown as individual problems.

MATH PROBLEM NO. 1

To find the **fourth dimension of a transition** joint when only three of the dimensions are given, as shown in Fig. 11-1.

Rule: Divide the **area** of the known opening, having two dimensions, by the dimension given for the opposite end.

Solution: 14″ x 12″ = 168 sq. in. = area of known end
168″ ÷ 8″ (dimension given for opposite end) = 21″
Opening at opposite end is then 21″ x 8″

Math Problems

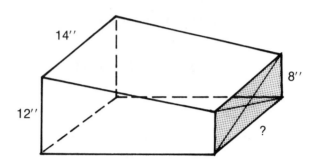

14''

8''

12''

?

Fig. 11-1

MATH PROBLEM NO. 2

To find the **dimensions of a square outlet** with the same area as that of a given **rectangular inlet.**

Rule: Find the area of the given *rectangular opening* and extract the *square root* of the product: this will be the dimension of the desired *square outlet.*

Solution: The rectangular opening is given as 8'' x 22''
22'' x 8'' = 176 sq. in.
The square root of 176 is 13.26 or 13-1/4 which will be the size of the square outlet, 13-1/4" x 13-1/4" having the same area.

NOTE: A chart giving **Squares of Mixed Numbers** *is invaluable in solving this type of problem.*

MATH PROBLEM NO. 3

To determine the amount of metal to be cut out of the circle, Fig. 11-2, which represents the flat pattern for a **stack cap.**

Rule: To find the width of **cut-out** C, multiply the difference between the diameters of the **base** and the **flat pattern** by 3.1416 (pi) or use the **Circumference Chart** (see **Index**).

Example: A is the elevation of the **cap,** Fig. 11-2, showing the required diameter to be 34''.
B is the radius of the flat pattern, which is 18-1/4''.

Solution: 18-1/4'' (radius of pattern) x 2 = 36-1/2'' = diameter
36-1/2'' − 34'' = 2-1/2'' = difference in diameters
The Circumference Chart shows that a circle of 2-1/2''

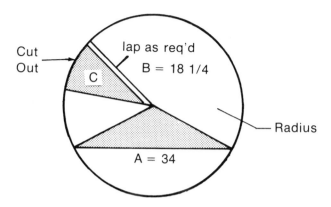

Fig. 11-2

has a circumference of 7.854 or 7-13/16″, which is the closest working fraction. This, then, is the arc to be cut out at *C* in Fig. 11-2 less allowance for lap.

The same result may be obtained by subtracting the *circumference* of the **base** from the *circumference* of the **flat pattern.**

MATH PROBLEM NO. 4

To find the **miter line** for an **elbow** of any number of **pieces** and of any **diameter.** Also, to find the **throat height** for any given **radius.**

Rule: Select the **factor** opposite the number of pieces required for the *elbow* and multiply this decimal by the *diameter* of the *elbow.* The answer will be the amount of **rise.**

Multiply this **same** factor by the required *radius* to determine the **throat height** of the *elbow.*

Discussion: The various factors shown here have been prepared in *chart* form as a convenience in preparing the *layout* for a *round elbow.* These decimal factors are derived from *trigonometry.* They are the *tangent functions* for various *degrees,* based upon the number of divisions, or segments, in the degree angle required.

While the most frequently used *degree angles* are shown, occasions may arise when a *special angle* is required. In this case, refer to the **tangent** *fuctions* in a Trigonometry Table to find the *factor* which applies to the special *degree.*

Math Problems

Example: Find the **rise** and **height** of the **throat** of a 7-piece *elbow* having a 12'' *radius* and a 9'' *diameter*.

Solution: Referring to the Chart, select the *decimal factor* opposite 7 pieces, which is .132. Multiply this by the *diameter* to determine the amount of **rise**. Using the same *decimal factor*, multiply it by the *radius* to determine the *throat height*, as shown in Fig. 11-3.

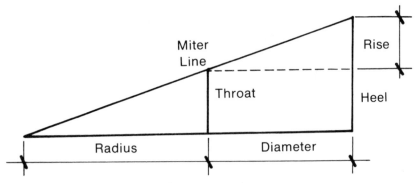

Fig. 11-3

ELBOW RISE CHART

No. Pieces or Gores	or Segs.	90 Degree End piece or segment	Factor	75 Degree End piece or segment	Factor	60 Degree End piece or segment	Factor
3	4	22-1/2°	.414	18-3/4°	.339	15°	.268
4	6	15°	.268	12-1/2°	.222	10°	.176
5	8	11-1/4°	.199	9-3/8°	.165	7-1/2°	.132
6	10	9°	.159	7-1/2°	.132	6°	.105
7	12	7-1/2°	.132	6-1/4°	.110	5°	.088
8	14	6-1/2°	.113	5-5/16°	.093		
9	16	5-5/8°	.098	4-11/16°	.083		
10	18	5°	.089				

		45 Degree		30 Degree		22-1/2 Degree	
3	4	11-1/4°	.201	7-1/2°	.132	5-5/8°	.099
4	6	7-1/2°	.132	5°	.088	3-1/2°	.061
5	8	5-5/8°	.099	3-3/4°	.066	2-7/8°	.050
6	10	4-1/2°	.078	3°	.052	2-1/2°	.044
7	12	3-3/4°	.066	2-1/2°	.044	1-7/8°	.033

MATH PROBLEM NO. 5

To determine the **length** of any one **side** of a **triangle** when the other two sides are known.

The *triangle* in Fig. 11-4 shows the **base,** *B*, the **altitude** or **height,** *A*, and the **hypotenuse,** *C*. For this example, the following values are assigned: $A = 3, B = 4, C = 5$.

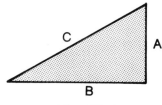

Fig. 11-4

Rule: To find *C* or the *hypotenuse*, add the *square* of *A* to the *square* of *B* and, from this **sum**, extract the *square root*.
$$C = \sqrt{A^2 + B^2}$$

Example: $3 \times 3 = 9$ $4 \times 4 = 16$ $9 + 16 = 25$
The *square root* of $25 = 5 = C$

Rule: To find *A* or the **altitude,** subtract the *square* of *B* from the *square* of *C* and, from this result, extract the *square root*.
$$A = \sqrt{C^2 - B^2}$$

Example E: $5 \times 5 = 25$ $4 \times 4 = 16$ $25 - 16 = 9$
The *square root* of $9 = 3 = A$

Rule: To find *B* or the **base,** subtract the *square* of *A* from the *square* of *C* and, from this **result,** extract the *square root*.
$$B = \sqrt{C^2 - A^2}$$

Example: $5 \times 5 = 25$ $3 \times 3 = 9$ $25 - 9 = 16$
The *square root* of $16 = 4 = B$

MATH PROBLEM NO. 6

To find the *throat* and *heel* **stretchout** or **length** when the width of the cheek and throat radius are known.

Rule: To find the **throat** stretchout, multiply the radius of the throat by 1.57 (this *factor* being one-half of *pi* or 3.1416).

Math Problems

For the **heel,** add the cheek width to the throat radius and multiply this by the same *factor, 1.57.*

Discussion: The above applies to 90° or a full quarter-circle. To determine the length or stretchout for *angles* under 90°, use the same procedures but apply the *factor* shown below for the angle required.

Degrees	Factor
90	1.570
60	1.050
45	0.785
30	0.524

Example: Find the *stretchout* of a 30° *elbow* with a throat radius of 12''.

1.57 x 12'' = 18.84 or 18-7/8''

Example: Find the *stretchout* of a 30° *elbow* with a throat radius of 12''.

(From chart) .524 x 12'' = 6.288 or 6-5/16''

MATH PROBLEM NO. 7

To determine the **unknown length of the throat** for a **tapin** when the given **pitch** is either longer or shorter than the width of the cheek.

Example: In Fig. 11-5, the *pitch* is given as 4'' in 38'', although the *cheek* of the *tapin* is only 24''. The heel is 10'' high. Find the rise at *r* and the height of the throat, *h.*

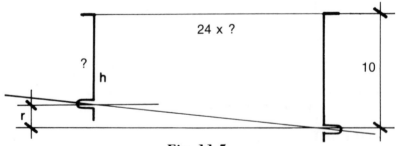

Fig. 11-5

Solution: First, determine the *rise* per inch and multiply this by the width of the cheek, or 24''. This will give the *rise* at *r* which is then subtracted from the heel height, 10'', to give the throat height, *h.*

302

$4.000'' \div 38'' = .105''$ rise per $1''$

$.105 \times 24'' = 2.520$ or $2\text{-}1/2'' = r =$ rise in $24''$

$10'' - 2\text{-}1/2'' = 7\text{-}1/2'' = h =$ throat height

MATH PROBLEM NO. 8

To find the **height** or **width** of the first *piece* and each remaining *piece* of an elbow, regardless of the **elbow size** or the **number** of **pieces.**

Rule: Multiply the number of *pieces* required by 2 and, from this product, subtract 2. The resulting answer is the number of spaces into which the throat or heel curve must be divided to establish the *miter line*. The end pieces being the first and last spaces, the remaining center *pieces* require two spaces each.

Examples: 3-piece elbow $3 \times 2 = 6$ $6 - 2 = 4$ (no. of divisions)
4-piece elbow $4 \times 2 = 8$ $8 - 2 = 6$
5-piece elbow $5 \times 2 = 10$ $10 - 2 = 8$

Discussion: To determine the height of the first *miter line*, divide $90°$ by the calculated number of divisions, as shown below. This gives the degree of the first *miter* and, using a *protractor* at the radius point, will establish the *miter*.

This process eliminates the need to draw out a full quarter-circle and then dividing it with *dividers*. It is also possible, using this *method*, to calculate each *miter*, by a given degree, as shown here.

Example: For a *3-piece elbow:*
$90° \div 4 = 22\text{-}1/2° =$ first miter

The following are the degrees at each *miter*
3-piece = $22\text{-}1/2°, 67\text{-}1/2°, 90°$
4-piece = $15°, 45°, 75°, 90°$
5-piece = $11\text{-}1/4°, 33\text{-}3/4°, 56\text{-}1/4°, 78\text{-}3/4°, 90°$
7-piece = $15°, 30°, 45°, 60°, 75°, 90°$

NOTE: The above method can be replaced by using the **Elbow Rise Chart** *shown elsewhere in this book (see* **Index**).

MATH PROBLEM NO. 9

To locate the **radius point** for scribing the *throat* and *heel* curves of a true proportioned **ogee** *offset* and to determine the *length* of its *wrappers.*

Math Problems

METHOD NO. 1

Rule: The square of the **offset** plus the *square* of the **length,** divided by 4 times the **offset,** equals the **radius** point from the center of the cheek.

Formula: $\dfrac{\text{off}^2 + L^2}{4 \times \text{off}}$

METHOD NO. 2

Rule: The **length** squared, divided by 4 times the **offset,** plus the **offset** divided by 4, equals the **radius point** from the center of the cheek.

Formula: $\dfrac{L^2}{4 \times \text{off}} + \dfrac{\text{off}}{4}$

NOTE: Either of the above formulas can be readily applied to locate the radius point. The cutter should adopt the one he finds easiest to remember, as this is the purpose in showing them both.

Rule: To determine the **length** of **wrappers,** add the *square* of the **offset** to the *square* of the **length** and, from this **sum,** extract the *square root.* Then, **subtract** the given **length** of the *fitting* and **divide** by 3. This amount, added to the *square root* already obtained, will be the **length** of the **wrappers; not** including the amount necessary for *lock* and *flange.*

Formula: $\dfrac{\sqrt{\text{off}^2 + L^2} - L}{3} + \sqrt{\text{off}^2 + L^2}$

Example: Using **method No. 1,** find the **radius** point for an *offset fitting* which is 46-1/4″ long with an offset of 14″.

$$\frac{(14 \times 14) + (46\text{-}1/4 \times 46\text{-}1/4)}{4 \times 14} =$$

$$\frac{196 + 2139.06}{56} = \frac{2335.06}{56} = 41.696 \text{ or } 41\text{-}11/16″$$

41-11/16″, the closest workable fraction, is the radius point from the center of the fitting.

Example: Using **method No. 2,** find the **radius** point for the *offset fitting* described in the previous example.

$$\frac{46\text{-}1/4 \times 46\text{-}1/4}{4 \times 14} + \frac{14}{4} =$$

304

Problem No. 10

$$\frac{2139.06}{56} + \frac{14}{4} = 38.197 + 3.5 = 41.697 \text{ or } 41\text{-}11/16''$$

Problem: Find the length of the **wrappers** for the *offset fitting* described in the previous examples.

$$\sqrt{\frac{(14 \times 14) + (46\text{-}1/4 \times 46\text{-}1/4) - 46\text{-}1/4}{3}} +$$

$$\sqrt{(14 \times 14) + (46\text{-}1/4 \times 46\text{-}1/4) =}$$

$$\sqrt{\frac{2335.06 - 46.25}{3}} + \sqrt{2335.06} = \frac{48.3125 - 46.25}{3} + 48.3125 =$$

$$\frac{2.0625}{3} + 48.3125 = .6875 + 48.3125 = 49'' = \text{the length}$$

of the wrappers **not** including the *lock* and *flange*.

MATH PROBLEM No. 10

To find the **ratio of given area** required for each **take-off.**

Example: As shown in Fig. 11-6:

$A = 12'' \times 12''$
$B = 14'' \times 12''$
$C = ? \quad \times 12''$
$D = ? \quad \times 12''$

The combined opening of C and D is given as $18'' \times 12''$. However, it must be determined how to split the $18''$ in proportion to the *outlets* A and B.

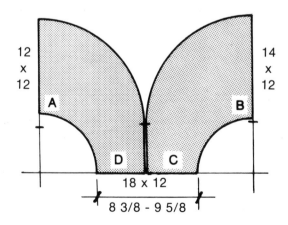

Fig. 11-6

305

Math Problems

STEP 1. Find the area of each *outlet*.

$12'' \times 12'' = 144$ sq. in. = area of A
$14'' \times 12'' = 168$ sq. in. = area of B
$18'' \times 12'' = 216$ sq. in. = area of C and D combined

STEP 2. Determine percentage of area A, then B, to the combined total of areas A and B, using the following method:

Add: $144'' + 168'' = 312''$
Divide:

$168''$ by $312'' = .538$ = ratio of B
$144''$ by $312'' = .462$ = ratio of A

Multiply:

$216'' \times .538 = 116.208$ = area of C
$216'' \times .462 = 99.792$ = area of D

Divide:

116.2 by $12'' = 9.666$ or $9\text{-}11/16''$ = side of C
99.79 by $12'' = 8.3158$ or $8\text{-}5/16''$ = side of D

For more workable figures in proportioning, use $9\text{-}5/8''$ and $8\text{-}3/8''$.

NOTE: 12 represents the 12'' side of the 18'' x 12'' opening, D *plus* C.

MATH PROBLEM No. 11

To find the distance between the *base* of the **frustrum of the cone** and the point of intersection at A, Fig. 11-7, or length K.

Rule: Subtract the **small** *diameter* from the **large** *diameter.* Multiply the **large** *diameter* by the height of the *fitting.* Divide this product by the difference of the two diameters just found. This will give the length of the center line to locate point A, as shown.

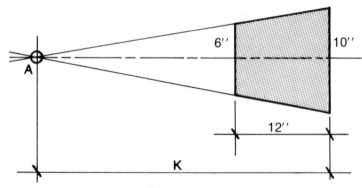

Fig. 11-7

306

Solution: $10'' - 6'' = 4''$ $12'' \times 10'' = 120''$
$120'' \div 4'' =$ length of center line to locate point A, which will be the point of intersection, or **apex.**

NOTE: When this method proves impractical, refer to **Tapers Having Slight Taper,** *shown elsewhere in this book (see* **Index** *for* **Round Layout Development).**

MATH PROBLEM NO. 12

Proportioning the Stack Cap

The *cap* is used extensively and can be made very easily. However, each *cap* should be proportioned by a set *rule*, with the **diameter** of the **stem,** D, being the determining *factor* for the *dimensions of* A, B, and C.

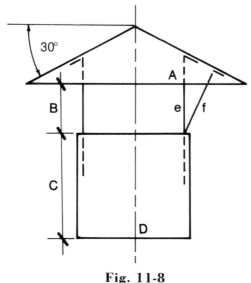

Fig. 11-8

The length of C can be a matter of choice, but A and B should be as follows:

$$A = 2 \times D \qquad B = 1/2 \text{ of } D$$

The *pitch* of the *cap* is 30 degrees.

The brackets, shown at e, can be straight up for the smaller *caps.* However, for the larger *caps*, they should be bent outwards, such as indicated by f, Fig. 11-8, to give greater support.

It is also advisable to use 4 brackets for a large *cap* having a *stem diameter,* D, of $13''$ or over. Also add a double bead around the *cap*, when possible, for added stiffness.

Math Problems

MATH PROBLEM NO. 13

To change a **decimal into a fraction** of **equal** or **nearly equal** value.

Discussion: Although most of our *mathematic problems* are calculated through the use of decimal figures, our final figure must be transformed into a **workable fraction.** This *fraction* is one that can be read on a *rule*, thereby making it possible to transfer this figure to the *layout*. While this process is either done mentally or taken from a *Decimal Table*, the *cutter* may find occasions where it is necessary to make this calculation mathematically. The *fraction* may be required in a given **denominator,** or merely reduced to a **common fraction.** Both *methods* are given here, as follows:

Rule 1: To reduce a *decimal* to a **common fraction,** place the number *1* under the **decimal figure** with as many ciphers at its right as there are *decimal* places. Then divide both figures, called the **numerator** and **denominator,** by the same number, until the lowest fraction is obtained.

Example: $.125 = \dfrac{125}{1000} = \dfrac{25}{200} = \dfrac{5}{40} = \dfrac{1''}{8}$

Rule 2: In reducing a *decimal* to a *fraction* having a required denominator, the *decimal* is multiplied by the given figure with the same figure being placed under the results.

Example: $.625$ in 8ths $= .625 \times \dfrac{8}{8} = \dfrac{5''}{8}$

When the **whole number** is also included, it is added to the answer as follows: 39.625 in 8ths becomes 39-5/8''.

MATH PROBLEM NO. 14

To determine the **amount of metal to be mitered** from a joint of *duct* or *pipe* to be used as an **offset.**

Rule: Multiply the width or height of the *pipe* by the *offset*. Then divide this product by the length.

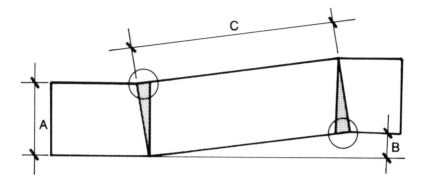

Fig. 11-9

Example: As shown in Fig. 11-9:

A = width or height	$A = 12''$
B = amount of offset	$B = 4''$
C = length of duct or pipe	$C = 36''$

$12'' \times 4'' = 48''$

$48'' \div 36'' = 1.33$ or $1\text{-}5/16''$ (.3125 being the nearest
workable figure)

Amount to be *mitered* from *duct* or *pipe* is $1\text{-}5/16''$.

*NOTE: This method of offset is often used on the job where the need
to offset a run of duct may occur and a regular offset
fitting is not available.*

MATH PROBLEM NO. 15

To determine the **degree of an angle,** without using a *protractor.*

METHOD A
Using a 4-foot Circumference (Bench) Rule.

This *method* is based upon the specific fact that, by scribing a
quarter-circle using a *radius* of 18-13/16'', as shown in Fig. 11-10,
the *circumference* edge of the *bench rule* will, when pulled around
this quarter circle, measure off exactly 90°.

The reading will accurately give each degree from 1 to 90, with
fractions thereof, on the *circumference scale* of the *rule*, which is
in 1/8'' divisions.

Math Problems

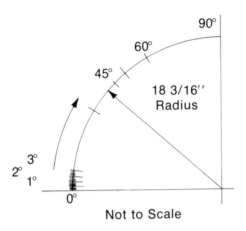

Not to Scale

Fig. 11-10

METHOD B
Using a Square as a Protractor.

Although this *method* is quite accurate for determining *angles* in 5° steps, it will not be dependable to attempt any in-between degrees such as offered by Method A.

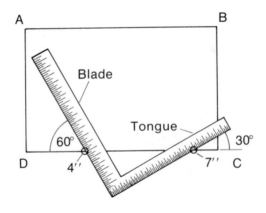

Fig. 11-11

To determine *angles* in 5° steps is a simple matter such as shown here in Fig. 11-11. The 7″ mark is used as a **set** number on the *tongue* of the *square*, while the *blade* is moved up or down to the figure given for the *blade* in the Chart. The desired *degree* of *angle* is shown opposite each figure, with the *angles* formed by *tongue* and *blade* always adding up to 90°. *A, B, C, D* represents the metal sheet.

Degree Tongue	Tongue	Blade	Degree Blade
5°	7″	9/16″	85°
10°	7″	1-1/4″	80°
15°	7″	1-7/8″	75°
20°	7″	2-9/16″	70°
25°	7″	3-3/16″	65°
30°	7″	4	60°
35°	7″	4-7/8″	55°
40°	7″	5-7/8″	50°
45°	7″	7	45°

When accuracy is required, such as in large *fitting layout* work, all *angles* should be determined by means of *trigonometry*. This is explained in the *trigonometry section* of this *book*, and the *cutter* should be thoroughly familiar with its application.

MATH PROBLEM NO. 16

To determine the **required straight duct between the two 45° angles,** knowing the amount of **offset.**

The development of this problem is based on a *factor (function value)* derived through *trigonometry*, which makes this *method* easily adaptable in actual *shop* problems.

The *function* used in this problem is referred to as the **versed sine** or **vers,** which is the abbreviated form more often encountered. The *function* is shown here in Fig. 11-12 at *V*, which is the span between the points *c-b*. The shaded area is merely to bring out more clearly the actual relationship of the *vers function* to the problem.

Example: Assume the problem shown in Fig. 11-12 to be of these proportions:

Math Problems

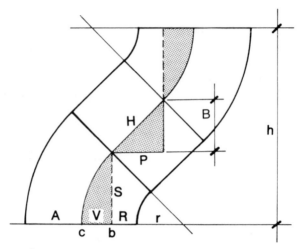

Fig. 11-12

cheek = 24'' (A = 1/2) & (R = 1/2)
r = 10'' radius
V = vers c-b (.29289 or .2929)
B = base of right triangle (side adj.)
P = perpendicular (side opp.)
H = hypotenuse
h = overall height (unknown)
S = sine function (.70711 *function value*)
offset = 24''

The development of the problem is figured from center of the *angles* as follows:

10'' rad. + 12'' (1/2 of 45° cheek)   10'' + 12'' = 22''
Using *factor* 0.2929:
22'' x 0.2929 = 6.4438 x 2 = 12.8876 or 12.89''
Width of offset = 24''
24''−12.89'' = 11.11'' or, as a workable fraction, use
11-1/8''.

This 11-1/8'' represents the *base* and *perpendicular* of the shaded *triangle* shown in the Fig. 11-12. It is obvious that the *hypotenuse* of this *triangle* is the length of *pipe* required to make the completed *offset.*

The *hypotenuse* can be determined in the same manner shown in **Math Problem No. 5.** However, a **plumber's rule,** which has a 45° scale on one side, is a quick means of obtaining a correct answer, which is 15-3/4'' for H.

The *reader* is referred to a similar problem shown as **Math Problem No. 21**, *Formula for Offset Connection.*

Although these two *methods* seem quite alike, there is one specific difference. In the *Formula for Offset Connection*, the development is to a pre-determined height, whereas the *method* shown here is developed to a specific amount of offset, but the height is **unknown.**

It is often necessary to know, before the development is made, just how long the finished *fitting* will be. Therefore, in order to determine the length, another calculation must be made, i. e., determining the height S, or the *sine function*. As this is explained in the section on Trigonometry, no further discussion will be given here other than the procedure for this problem.

$12'' + 10'' = 22''$ (represents the radius here)
$22'' \times .70711 = 15.55642''$ or $15\text{-}9/16''$
$15\text{-}9/16'' \times 2 = 31\text{-}1/8'' + 11\text{-}1/8'' = 42\text{-}1/4'' = h$
Overall length, less flanges $= 42\text{-}1/4''$

Although this is basically used for the 45° *offset*, it is possible to apply the same *method* to other *angles* by using the following *functions* taken from a **Versed Sine** *Chart:*

Degree	Versed Sine	Sine
60°	.5	.866
45°	.29289	.70711
30°	.134	.5
22-1/2°	.07612	.3827
15°	.03407	.2588

For angles other than 45°, the *triangle* forms in the same manner, but with each leg of a different length. This requires determining of the *hypotenuse* by dividing the side opposite, P, by the *sine function* appertaining to given *degree* of *angle*. If *base B* or *side* adjacent is required, apply the *method* described in *Trigonometry* (see **Index**).

MATH PROBLEM NO. 17

Finding the **size** and **speed** of **pulleys** and **belts.**

By means of the *method* shown here, the speed or diameter of either the **driven** or **driving** pulley may be determined.

The **driving** pulley described here is the one on the shaft of the *motor*, with the **driven** pulley being on the *blower.*

Math Problems

Problem A: Find the **speed of the driven pulley,** knowing the diameter and speed of driving pulley and diameter of the driven pulley.

Rule: Multiply the speed of the **driving** pulley by its diameter, then divide this by the diameter of the **driven** pulley. The answer will be the driven pulley's speed in RPM.

Problem B: Find the **diameter of the driven pulley,** knowing the diameter and speed of driving pulley and speed of the driven pulley.

Rule: Multiply the diameter of the driving pulley by its speed, dividing this by the required driven speed. The answer will be diameter of the driven pulley in inches.

Problem C: Find the **diameter of the driving pulley,** knowing desired speed, the diameter of the driven pulley and RPM of the motor.

Rule: Multiply the diameter of the driven pulley by its speed, then divide this by the speed of the driver pulley. The answer will be the diameter in inches.

Problem D: Find the **speed of the driving pulley,** knowing the diameters of both pulleys and the speed of the driven pulley.

Rule: Multiply the diameter of the driven pulley by its speed, then divide this by the diameter of the driving pulley. This will determine the speed of the driving pulley in RPM.

Problem E: Find the **diameter** necessary to produce any **required number of revolutions.**

Rule: Multiply the diameter of the pulley whose speed is known by its revolutions. Then divide by the revolutions at which the other pulley is to run.

Problem F: To find the **circumference of a pulley.**

Rule: Multiply diameter by 3.1416 (pi) or 3-1/7.

MATH PROBLEM NO. 18

Determining the **length of belts.**

Rule: Add the diameters of both pulleys in inches and multiply sum by 1.57 (1/2 of 3-1/7). To this product, add twice the distance between *shaft* centers, for the length.

314

NOTE: Although it is usually possible to reorder belts from the number given on the old belt, it is necessary to be certain that the correct belt has been in operation.

Length alone is not sufficient in selecting a V-belt, as they are required to fit a sheave or pulley at a certain depth.

It is also inadvisable to select belts of too many plies or heavier than necessary, in order to obtain best service and economy.

Determining the **speed of belts.**

Rule: Multiply the circumference of a driving pulley in feet, by number of revolutions per minute to obtain the belt speed in FPM.

MATH PROBLEM NO. 19

To **divide a circle** into an equal number of divisions, using a given **factor.** This *factor* is derived from a *trigonometric function.*

Multiply the diameter of the circle in question by the given **factor,** this product being the length of chord C, Fig. 11-13. Using *dividers*, this span can then be stepped off around the circumference, resulting in the number of equidistant divisions required.

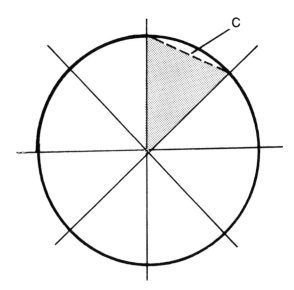

Fig. 11-13

Math Problems

Number Divisions	Factor	Number Divisions	Factor
3	.8660254	14	.2225208
4	.7071067	15	.2079116
5	.5877865	16	.1950903
6	.5000000	17	.1837495
7	.4338837	18	.1736481
8	.3826834	19	.1645945
9	.3420201	20	.1564344
10	.3090170	21	.1490422
11	.2817325	22	.1423148
12	.2588190	23	.1361666
13	.2393157	24	.1305262

MATH PROBLEM NO. 20

Determining the exact **stretchout** required to develop **cylindrical items,** such as *boiler breeching, band iron rings, tanks,* etc., to a given ID or OD measurement, by means of the chart shown here.

GAUGE	FACTOR	GAUGE	FACTOR	GAUGE	FACTOR
3/8	1.178	1/8	.392	20	.117
5/16	.981	12	.343	22	.098
1/4	.785	14	.248	2⁴	.078
3/16	.589	16	.198	26	.058
10	.442	18	.157	28	.049

The *factors* shown are the difference between each of the three *diameters* which develop when rolling heavy metal. The *factor* is selected according to the gauge metal being used and is applied as follows:

For **inside diameter, add** the *factor.*

For **outside diameter, subtract** the *factor.*

Example: To obtain an **inside** diameter of 20″, using 1/4″ plate, **add** the *factor* of .785 to the *circumference* of 62.83″. This makes a *stretchout* of 63.61, the *circumference* of a 20-1/4″ diameter circle.

To obtain an **outside** diameter of 20″, using 1/4″ plate, **subtract** the *factor* of .785 from the *circumference* of 62.83″. This *stretchout* of 62.05″ is the *circumference* of a 19-3/4″ diameter circle. As each figure represents the **mean** diameter, it is obvious that to obtain **inside** diameter the **mean** diameter must be lengthened out, and to obtain an **outside** diameter the **mean** diameter must be shortened.

NOTE: 3/16″ and over is regarded here as **plate.**

MATH PROBLEM NO. 21

Formula for Offset Connection

Although this **formula** can be used for *piping layout* in general, it will be found invaluable for determining the length of both *round pipe* and *square* or *rectangular duct* required between two **angles** of **any** degree and to **any** amount of *offset.*

Since *trigonometry* is the *method* of development applied here, decimals will be used throughout, with the final calculation being changed to the closest workable fraction.

Shown in Fig. 11-14 is the *layout* of the *offset*, plus various letters representing the following meanings in the formula:

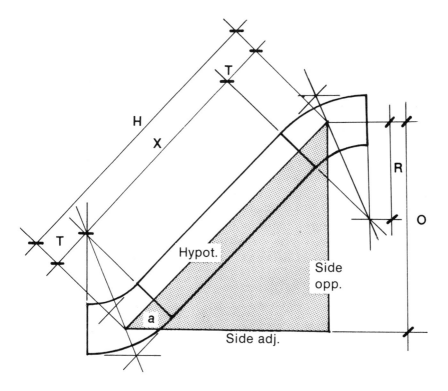

Fig. 11-14

Math Problems

a = Angle of Offset
O = Amount of Offset
R = Center Line radius
T = Center Line/Half of Angle (Tangent for 1/2 Angle)
H = Diagonal Point-to-point Stretchout or Hypotenuse
D = Diameter of Pipe and Angles
X = Length of Duct or Round Pipe Required
C = Cheek Width of Duct

Formula: (O x csc a) -2 (R x tan 1/2 a)

Example: Offset = 24''   Offset Angle = 45°
Center Line Radius = 15''   Pipe Diameter = 6''
(24 × csc 45°) -2 (15 × tan 22-1/2°) =
(24 × 1.414) -2 (15 × .414) = (33.936) -2 (6.210) =
33.936 $-$ 12.42 = 21.516 or 21-1/2''

NOTE: The cosecant function is for the degree angle being used. However with the tangent function, the angle is only being calculated to the center. Therefore, determine one-half of the angle's degree and use the tangent for that degree.

To determine the finished height of the *offset fitting*, having calculated the *point-to-point* length or *hypotenuse*, as shown in Fig. 11-14, use the *cosine* for the *angle* at *a* and multiply by the *hypotenuse* to obtain the length of side *adjacent* which, when added to 2 x T, will be the height.

Example: 33.936 or 34 x .70711 = 24.04174*
24.04 + 12.42 = 36.46 or 37-1/2'' total height
*In all 45 degree *angles* both legs of the *triangle* are the same length, requiring no calculating as was shown here. By knowing the amount of *offset*, we also have this length. The calculations made here were merely for explanatory purposes as other problems having different degree *angles*, will require this *calculating*.

Degree	Cosecant	Degree	Tangent
75°	1.035	37-1/2°	.767
60°	1.154	30°	.577
45°	1.414	22-1/2°	.414
30°	2.000	15°	.268
15°	3.864	7-1/2°	.132
10°	5.759	5°	.087

chapter XII

GEOMETRY PROBLEMS

Fig. 12-1
LOCATING THE EXACT CENTER OF A CIRCLE, WHEN THE CIRCUMFERENCE IS GIVEN

Let a,b,c represent points on the given circle. From point b on the circle, using any radius, scribe arc *1-2*. Using the same radius from points a and c, scribe the intersecting arcs *3-4* and *5-6*. By drawing lines *7-8* and *9-10* through these intersections, the two lines will cross at the exact center of the circle, point x.

Fig. 12-2
TO BISECT AN ANGLE

With J, K, L as the given angle, use K as center and scribe an arc of any convenient radius. This establishes points M and N. With the compass set at a radius of more than one-half the distance from M to N, use these points as centers and scribe intersecting arcs as shown at O. A line that is drawn through O and K bisects the angle.

Fig. 12-3
INSCRIBING A CIRCLE IN A GIVEN TRIANGLE.

Draw the triangle as shown, using the given dimensions, and bisect the angles *e-f-d* and *f-d-e* with arcs intersecting at g and h.

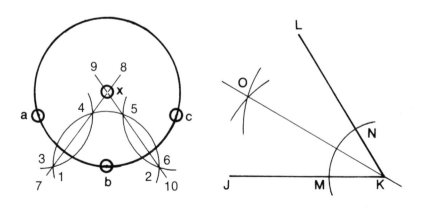

Fig. 12-1   Fig. 12-2

Geometry Problems

By drawing straight lines through points *f-g* and *h-d*, the exact center, or radius point *i* for the circle, is located.

Fig. 12-4
INSCRIBING AN EQUILATERAL TRIANGLE (EQUAL SIDES) IN A GIVEN CIRCLE.

The diameter of the given circle is the length of line *j-k*. Using the radius of the given circle, with *k* as center, scribe arc *m-n*. By connecting points *j-m-n* with straight lines, the triangle is completed.

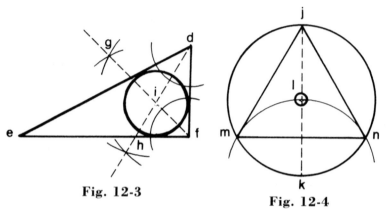

Fig. 12-3

Fig. 12-4

Fig. 12-5
TO DESCRIBE THE SEGMENT OF A CIRCLE OF ANY GIVEN CHORD OR HEIGHT

Draw the line *A-B* as given chord. Place the perpendicular line *M-N* at any random length, with *P-M* being the given *height*. Connect *M-B* with a straight line and bisect this line as shown at *C-G*. Carry this line down across the perpendicular line *M-N* to *R* to establish the *radius* point from which segment *A-M-B* is drawn.

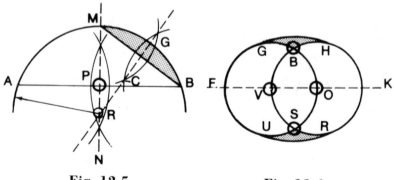

Fig. 12-5

Fig. 12-6

320

Fig. 12-6
FORMING AN OVAL OR ELLIPSE WITH A COMPASS

Draw the horizontal line *F-K* the length of the oval required. Set the dividers at one-third the major diameter, *F-K*, thereby locating points *V* and *O*. With these points as centers, describe the circles as shown. These two circles intersect at points *B* and *S*. Using these points as centers and with the compass set equal to the diameter of one of the circles, scribe the arcs *G, H* and *U, R* which complete the oval.

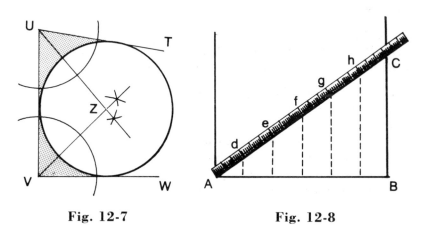

Fig. 12-7   Fig. 12-8

Fig. 12-7
CONSTRUCTING A CIRCLE TANGENT TO THREE INTERSECTING LINES

Given the three lines *T-U, U-V, V-W*, bisect the angles *T-U-V* and *U-V-W* by lines intersecting at point *Z*. *Z* will be the center of the required circle.

Fig. 12-8
DIVIDING A LINE, OF GIVEN LENGTH, INTO A SPECIFIED NUMBER OF DIVISIONS OR SPACES

Assume that the given line *A-B* is 28-1/2″ long and is to be divided into six parts. From point *B*, draw a perpendicular line of random length. Mentally determine what number larger than 28-1/2 would use six as a common denominator, which is, of course, 30. Place a bench rule at *A*, with the 28-1/2″ mark resting on *B*. Firmly hold the rule at *A* with the left hand, while the other hand raises the rule up the line *B-C* until the 30″ mark rests directly on the line. Draw a line across the rule from *A-C* and mark off each five inches,

321

Geometry Problems

shown at *d, e, f, g,* and *h*. Using a square up from *A-B*, carry all points down to the line *A-B*, which will be equally divided into six parts.

Fig. 12-9
CONSTRUCTING AN ANGLE SIMILAR TO GIVEN ONE

The given angle is shown by *T,R,S*. Using *R* as center, with any radius, scribe the arc *1-2* through both lines of the angle. For the similar angle, place line *V-W* the same as *R-S* and also scribe the same arc, using *V* as center. With *1* to *2* as the radius and *4* as center, intersect arc *3-4* at *5*, thereby establishing point *X*. A line from *V* through *X* completes the angle.

Fig. 12-10
CONSTRUCTING AN IRREGULAR FIGURE OF THE SAME SIZE AND SHAPE AS A GIVEN ONE

Draw line *2-3* equal in length to the line *E-F* in the given figure *D,E,F,G*. Scribe an arc of random radius from points *E, F, 2* and *3*. Set the dividers to radius *F-f* and, from point *3*, intersect the arc at point *5*. Draw a line through points *2* and *5*. On this line scribe an arc equal to radius *E-D* to establish point *1*. Repeat the procedure from points *E* and *2* to establish point *6* and, subsequently point *4*. To complete the figure, connect points *1* and *4* with a straight line.

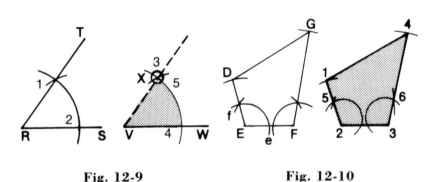

Fig. 12-9   Fig. 12-10

Fig. 12-11
CONSTRUCTING REGULAR PENTAGON INSIDE A GIVEN CIRCLE

With *H* as center, scribe the circle *I,J,K,L*. Draw the top diameters at *I-K* and *J-L* perpendicular to each other. Bisect the radius *H-I* by the line passing through *H-I* at *1*. With *1* as center, and

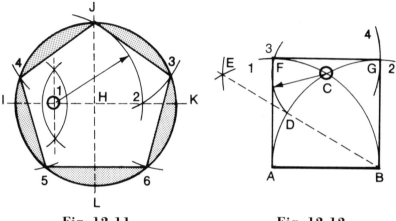

| Fig. 12-11 | Fig. 12-12 |

1-J as radius, scribe arc *J-2*. Using the same radius, and with *J* as center, scribe the arcs *3* and *4*. With *3* and *4* as centers, using same radius locate points *5* and *6*. By connecting these points, the pentagon is completed.

Fig. 12-12
CONSTRUCTING A SQUARE FROM A GIVEN SIDE

Line *A-B* represents the given side. Using this length as a radius, with *A* as center, scribe the arc *B-1* beyond the point of radius. Following the same procedure, only using *B* as center, scribe arc *A-2* which will intersect at point *C*. Bisect arc *A-C* at *D* by intersecting arcs at *E*. Using *C* as center, and *C-D* as radius, scribe arcs *3* and *4*, which will intersect arcs *1* and *2* at *F* and *G*. Complete the square by connecting the points of each corner.

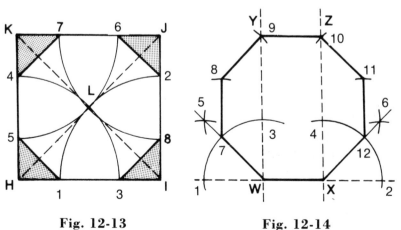

| Fig. 12-13 | Fig. 12-14 |

Geometry Problems

Fig. 12-13
TO INSCRIBE AN OCTAGON WITHIN A GIVEN SQUARE

Draw the given square to size and place diagonal lines from corner to corner to locate the center point *L*. With the radius equal to either corner, and each corner as center, scribe an arc crossing the square at points *1,3,8,2,6,7,4,5*. To complete the octagon, join these points with straight lines.

Fig. 12-14
TO CONSTRUCT AN OCTAGON ON A GIVEN LINE

The given line, *W-X*, should be extended out on both sides as shown. Erect perpendicular lines upward from points *W* and *X* as shown, *W-Y* and *X-Z*. Set *trammels* to the width *W-X* and, with *W* and *X* as the radius points, scribe two quarter circles which locate points *1, 3, 4, 2*.

With *trammels* still set at the same setting, from points *1* and *3* then from *4* and *2*, scribe intersecting arcs as shown at *5* and *6*. Using a straightedge from points *W* and *X* to points *5* and *6*, draw in the lines to establish points *7* and *12*. With the same setting on the *trammels*, scribe short arcs from points *7* and *12* towards the top.

Set the *trammels* to the width of points *7-X* and, with point *X* as radius point, locate point *11*. Apply the same procedure from point *W* to locate point *8*. With *trammels* set the width of *W-X* and with points *8* and *11* as radius points, scribe short arcs to cross the perpendicular lines *W-Y* and *X-Z* locating points to complete the *octagon*.

Fig. 12-15
TO DRAW A CIRCLE THROUGH ANY 3 GIVEN POINTS

The given points are shown at *O,P,Q*, and are to be connected together with straight lines. These lines are bisected, as shown, and straight lines are carried through each until they meet at *T*, with *T* being the exact center. Thus from *T*, and with *Q,P,O* as radius, the circle can be drawn in.

Fig. 12-16
TO DEVELOP MITER LINE FROM A GIVEN ANGLE
(Bisecting an Angle)

Angle is *P, Q, R*. Using any radius as a center, scribe arcs *1* and *2*. With same radius at *1* and *2*, scribe arcs at *3*. Dash line is miter line. Arcs from *1, Q, 2* will locate parallel lines.

324

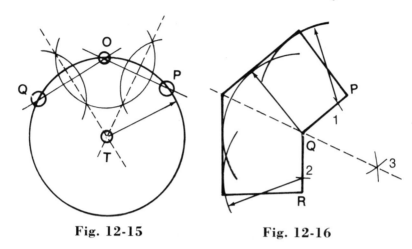

Fig. 12-15   **Fig. 12-16**

Although similar to Fig. 12-2, this method is applied to develop the cutting line between two pieces of angle iron, etc.

Fig. 12-17
TO CONSTRUCT A HEXAGON FROM A GIVEN SIDE

Describe a circle with radius equal to the given side *S-T*, and then carry a line across to establish the diameter *T-U*. Using the points *T* and *U* as centers, and the same radius, locate the points *1,2,3,4*. To complete the hexagon, connect all the points.

Fig. 12-18
TO DRAW AN OCTAGON WITHIN A GIVEN CIRCLE

Draw the given circle as required. Across the center of the circle, draw lines *W-X* and *Y-Z* perpendicular to each other, crossing the

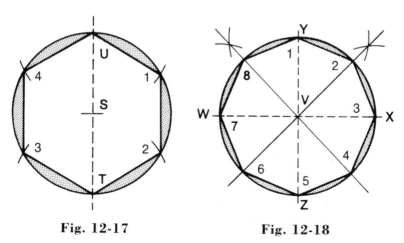

Fig. 12-17   **Fig. 12-18**

Geometry Problems

circumference at points *1*, *5*, *7* and *3*. Bisect the angles *Y-V-W* and *Y-V-X* and carry these points through the circumference at *2* and *8*. Then, by connecting with a straight line through point *V*, the points *6* and *4* are also found. To complete the octagon, connect all outside points on the circumference with straight lines.

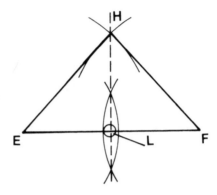

Fig. 12-19

Fig. 12-19
DRAW AN EQUILATERAL TRIANGLE FROM A GIVEN SIDE.

The line *E-F* being the given line, use its length as a radius and, with *E* and *F* as centers, scribe intersecting arcs at *H*. Connect with straight lines to form the triangle. Shown here: by bisecting line *E-F*, as indicated at *L*, the triangle is easily split in two.

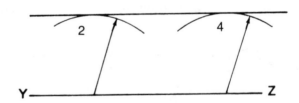

Fig. 12-20

Fig. 12-20
PARALLEL LINE TO A GIVEN ONE

Line *Y-Z* represents the given line. From any points along the line, and with the dividers set equal to the desired spacing from line *Y-Z*, scribe two arcs as shown at *2* and *4*. By holding a straight-edge against these two arcs, the parallel line is thus obtained.

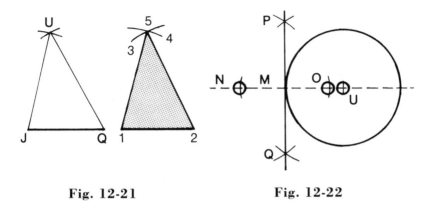

Fig. 12-21   Fig. 12-22

Fig. 12-21
TO COPY A TRIANGLE

Line *J-Q* is the base of the given triangle *J-Q-U*. Using this line as a radius, scribe the distance *1-2*, which becomes the base of the copy. Set the dividers equal to the lines *J-U* and *Q-U*. Intersect the arcs *3* and *4* to establish point *5*. To complete the triangle use connecting lines. This same procedure can be worked from any one of the other lines.

Fig. 12-22
DRAWING A TANGENT TO A GIVEN CIRCLE AT A GIVEN POINT ON ITS CIRCUMFERENCE

Draw a line through the center of the circle *U* and given point *M*. With *M* as center, describe arcs of equal radius, cutting this line at *N* and *O*. With *N* and *O* consecutively as centers, describe arcs of equal radius intersecting at *P* and *Q*. A line drawn through *P* and *Q* will be the required tangent.

Fig. 12-23
LOCATING OPPOSITE POINTS OF CIRCLE, BY MEANS OF A SQUARE

Although this has been included in **Geometry Problems**, it will be found to actually be a workable shop method for locating directly opposite sides of a round object. This could apply to either a pail, or possibly a round piece of pipe, which requires a damper. By closely observing Fig. 12-23, it will be noted that with the point of the *square* set on the circle, regardless of where the arms of the *square* are turned, they automatically locate the directly opposite points.

Geometry Problems

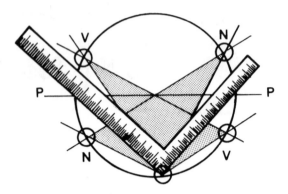

Fig. 12-23

NOTE: All problems of geometry have not been included here, since some are seldom, if ever, used in our work. Those that are shown and described here, are the ones most frequently used in sheet metal pattern drafting and should be at the cutter's finger-tips at all times.

Most sheet metal workers are familar with these geometry problems, having had them at one time or another. However, when not used often enough, they are easy to forget.

*For that reason, the writer feels they are a most important part of **Ready Reference**, so they can be referred to when necessary.*

It is well to fully understand all the problems shown. Although not used regularly, it should be possible for the cutter to recognize the occasions where they can be applied in layout work. This is important, as all layouts include, in one way or another, one or more geometry problems in their development.

chapter XIII

GENERAL INFORMATION

The purpose of this chapter is to provide a means whereby miscellaneous information could be grouped together. While not necessarily of vital importance to the *cutter* in *air conditioning*, this information should prove to be informative, helpful and interesting to the Reader. The material contained herein is not meant to be of a technical nature, but a mixture of *ideas*, *facts*, *tricks of the trade*, *terminology*, etc., which are in some way associated with our trade.

HOW TO FIND

AREAS

1. Area of a Circle
- **a.** Square of diameter x .7854
- **b.** Square of circumference x .07598
- **c.** Square of radius x 3.1416

2. Area of a Triangle
Base x 1/2 altitude

3. Area of a Parallelogram
Base x altitude

4. Area of a Rectangle
Length x width

5. Area of an Oval or Ellipse
- **a.** Multiply diameters together and the product x .7854
- **b.** 1/2 major axis x 1/2 minor axis x 3.1416

6. Area of a Circular Ring
Multiply sum of inside and outside diameters x their difference x .7854

7. Area of an Octagon
Square of diameter of inscribed circle x .828

8. Area of a Trapezoid
Altitude x 1/2 sum of parallel sides

9. Area of a Hexagon
Square of diameter of inscribed circle x .866

329

General Information

10. Area of a Sector of a Circle
Multiply the arc of the sector x 1/2 the radius

11. Area of Segment of a Circle
Determine area of sector having same arc; also area of the triangle formed by the radii and chord. Take the sum of these areas if the segment is greater than 180°, take the difference if less than 180°.

12. Area of Circumscribed Square
Area of the circle x 1.2732

13. Area of Inscribed Square
Area of the circle x .63662

14. Area of Cylindrical Surface
Circumference x length plus area of both ends

15. Area of a Pyramid or Cone
Diameter of base x 3.1416 x 1/2 the slant height

16. Area of the Frustum of a Cone (Slant surface only)
Multiply the sum of both circumferences x 1/2 the slant height

17. Area of a Sphere or Ball
a. Square of diameter x 3.1416
b. Diameter x circumference

SOLIDS (also referred to as **contents, volumes, capacity** or **solidity**)

18. Volume of a Cone (cubic inches)
Area of base x 1/3 the altitude.

19. Volume of a Sphere or Ball (cubic inches)
a. Cube of diameter x .5236
b. Cube of radius x 4.1888
c. Surface area x 1/6 of diameter.

20. Volume of a Cylinder (cubic inches)
Area of base x the length (cubic inches divided by 1728 = cubic feet)

21. Volume of a Cylindrical Ring
Add body diameter of ring to inner diameter of ring, multiply this sum by square of body diameter x 2.4674

22. Volume of a Rectangular Wedge
Length x breadth x 1/2 the height.

23. Volume of Segment of a Sphere
Base radius squared x 3 plus height squared x height x .5236

24. Capacity of Cylinder or Tank
Square of diameter x length x .0034 (gallons)
Square of diameter x length x .7854 (cubic inches)

25. Length of an Arc
Number of degrees x .017453 radius

26. To Obtain Square Feet
Square inches x .00695

27. To Obtain Cubic Feet
Cubic inches x .00058

28. Length of an Equal Cylinder
Diameter of a ball or sphere x .667

29. Diameter of a Circle
  a. Radius x 2
  b. Circumference divided by 3.1416
  c. Circumference x .31831
  d. Square root of the area x 1.128

30. Contents of a Square Tank in Gallons
Area of bottom x height (determines cu. ft.) x 7.48

31. Contents of a Rectangular Tank in Gallons
Length x width x depth (in inches) divided by 231

32. Equivalent Circles to a Known Square
  a. A side multiplied by 3.547 equals circumference
  b. A side multiplied by 1.1284 equals diameter
  c. A side multiplied by 4.443 equals circumference of its circumscribing circle
  d. A side multiplied by 1.414 equals diameter of circum-scribed circle
  e. Perimeter of square x .86623 equals circumference of equal square.

33. Side of a Square of Equal Area.
  a. Diameter x .8862
  b. Divide diameter by 1.1284
  c. Circumference x .2821
  d. Divide circumference by 3.545

34. Side of an Inscribed Cube
Radius of sphere x 1.1547

General Information

35. Side of an Inscribed Square.
    a. Diameter x .70711
    b. Circumference x .2251
    c. Divide circumference by 4.4428

SCRIBERS

Due to the fact that the *cutter* blocks out as much material as possible at the *shear*, from a prepared blockout sheet which includes allowances for various required edges, the greatest majority of his *layout* work is then a matter of working off the edges of the material.

For this purpose, his most important tool is the *scriber* and it is a matter of choice as to which type the *cutter* prefers to use. Some prefer the use of an adjustable *square* and *scratch awl* which they reset according to the size required.

It is, however, decidedly easier to use the type shown here and, since they are easily made from scrap pieces of stainless steel, the *cutter* can make a set of the sizes which he most frequently uses.

The *scribers* shown here are two different types: one being for use with the metal lying flat on the bench and the other for use with the metal either held up off the bench or off the edge of the bench itself. The writer prefers the first type which can be used with the metal being flat on the bench. This saves considerable handling of the material (wasted motion) and is somewhat more accurate when used in this manner.

The types used on the bench are shown in *A, B, C* and *E* of Fig. 13-1. Shown at *D* is the off-the-bench type which, like the others shown here, can be made in various sizes.

The scriber shown in *E* is perhaps the most useful of all, even though it is not used as frequently as the others. This *scriber* should be made in several sizes, such as 1/4", 1/2", 7/8", 1", and for the width used for the *pittsburg* edge.

This type of *scriber* is for use across the *sheet* when laying out such types of *fittings* as *straight transitions*, etc. The *scriber* is used by holding it against a straightedge or *rule* which is held from point to point across the sheet.

Although any desired size can be made, it will be found useful to have an adjustable type *scriber*, which can be purchased at most any hardware store.

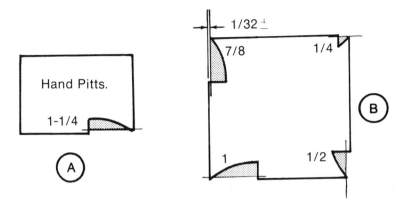

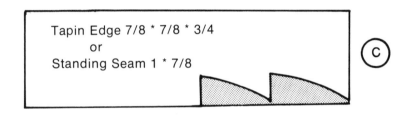

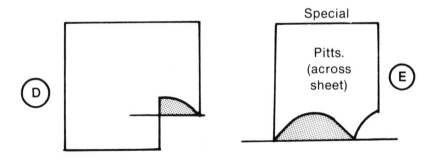

Fig. 13-1

General Information

TIME SAVER PATTERNS

Shown in Fig. 13-2 are the four most frequently used patterns, each representing a corner of the following items:

A. **Patch type door** (add 5-1/8'' to size)

B. **Cap/blank end** (add 3-3/8'' to size)

C. **Cut-off corner** (square elbow) (3'' / 3'' @ 45° = 4-1/4'' or 4'' / 4''
  @ 45° = 5-5/8'')

D. **Mitered pittsburg corner**

E. Blow-up of the notching at the corner intersection of two *pittsburgs, shown in D.*

Also shown are the following items:

F. **Rivet hole spacing template** (make several sizes)

G. Contains the following items:

        a. *S*, with stiffener (when formed)

        b. *S*, pattern for above

        c. *S*, flat (when formed)

        d. *D*, drive (when formed)

        e. *D*, pattern for above

H. Pattern for **government lock,** also referred to as a **clip** and **pocket lock:** standard size, to accept 7/8'' flange.

I. Same as above but for 1-3/8'' flange.

J. **Government lock** (when formed)

NOTE: The cutter will find it most useful to make similar patterns of such common items as door frame corners, louver blades, etc.

GENERAL INFORMATION

1. Rectangular duct should be made as square as possible and, unless unavoidable, the duct should not be more than four times as wide as it is deep.

2. When reducing size or changing shape of a duct, the angle of the slope should never be greater than 15° or 20°.

3. Doubling the diameter of a circle increases its area four times; by the same token, if the diameter of a pipe is doubled, its capacity also increases four times.

4. Double riveting is considered 16% to 20% stronger than single.

5. It is easier and more accurate to lay out **rivet** holes directly on the miter line, allowing lap on both sides of the connecting pieces.

334

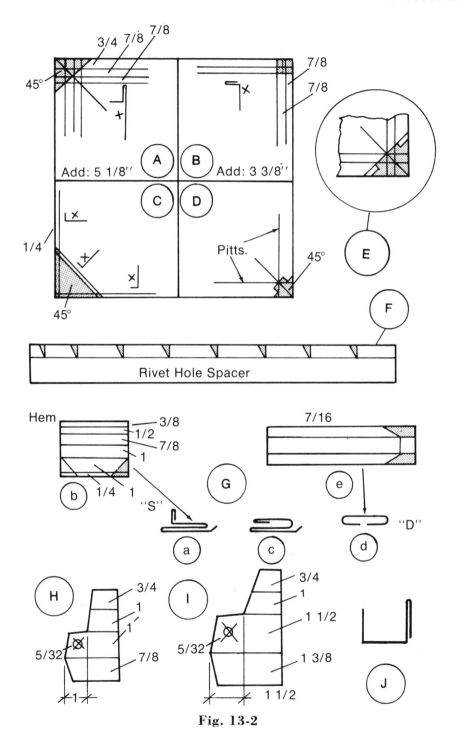

Fig. 13-2

General Information

6. Use 1-1/2 times *grille* size for header on end of duct run, i.e., the last *outlet*.

7. Keep in mind that all fittings are **rising** or **dropping,** either up or down, in the direction of the **air flow** which is towards the government lock.

8. For general use other than blowpipe work, a 3- or 5-piece elbow is most commonly used.

9. **Gross** weight of an article is the total weight including the package, crate or container, whereas, **net** weight is the actual weight of the article or contents only.

10. Air velocity (fpm) is equal to the volume (cfm) divided by the duct area in square feet.

ROUGHLY ESTIMATING METAL WEIGHT

There are occasions when a rough estimate is required regarding the weight of a certain piece of metal. It does not necessarily have to be exact; however, it must be reasonably close. The following figures are not exact weights, but are given in easy-to-remember fractions, which are slightly under the correct weights. For exact weights, refer to the Metals Table included in the Chart Section of this book.

Shown here is the approximate weight per square foot and the weight for a 10-foot *galvanized* sheet, 48 x 120, which is the most commonly used size.

Gauge	Lbs. per Sq. ft.	Lbs. per sheet
10	5-3/4	230
12	4-1/2	180
14	3-1/4	130
16	2-5/8	105
18	2-1/8	85
20	1-5/8	65
22	1-3/8	55
24	1-1/8	45
26	7/8	35
28	3/4	30
30	5/8	25

Soldering Tips

HELPFUL SOLDERING TIPS

It is the practice in many shops to have the *cutter* follow through the entire process of *layout, fabrication* and **soldering** when making *roof jacks, pans* and any similar items. The *cutter* must be capable of turning out a neat, water-tight soldering job.

Soldering is not difficult; however, it does require practice and the application of several simple *rules* based upon the type of metal being used and the gauge of the material.

1. When soldering *copper*, use a *hot* iron.

2. For *lead*, a comparatively *cold* iron is required, to keep the *lead* from developing holes or *running away*. Aluminum flux works quite well for this type of material.

3. For all soldering jobs, keep the irons in good shape, controlling the heat so that the irons do not become *red* or burn up.

4. When salammoniac is not available, substitute a piece of galvanized iron and acid as a means of tinning the irons.

5. To prepare *cut* acid, add small pieces of galvanized iron to *muriatic acid*. The *acid* in this form is used for soldering *copper*.

6. When soldering galvanized iron, some mechanics add a bit of water to the acid. The purpose of this is to weaken the acid a little, which makes it easier to clean when finished. Also, it helps considerably in checking the fumes and smoke which arise when being applied.

7. One of the best means of cleaning up after completing the *soldering* job is to use water with *baking soda* added. Applied with about a 2″ paint brush, this works far more efficiently than wiping with a rag.

TEMPLATE FOR HOLE LAYOUT

When making up a template for hole layout, accuracy can be assured by preparing it in the following manner: After laying out the spacing for the holes, prick mark and go over again, using a sharp center punch, which will set up a deep burr on the underside of the pattern or template. Then, use a sanding machine to remove these burrs, leaving a flat surface. This pattern will remain this same way

General Information

for considerably more use than otherwise and, at the same time, will not have a tendency to shift sideways or lift away from the angle frame or material being marked.

After a period of frequent use, the *pattern* will become high again and should be sanded again. The hole will become slightly larger but, by using a properly angled center punch, it will seat into the hole without being off center.

When having to drill any large size holes through heavy material, such as *angle iron*, *plate*, etc., always use a small size drill for a pilot hole. Then, being absolutely certain the material being drilled is securely clamped in place, drill the larger holes as required. Be certain also to use *cutting oil* in order to keep the *drill bit* from burning up, thereby retaining its sharp cutting edge for a longer period of time.

ABBREVIATIONS

The following *abbreviations* are most frequently encountered:

B. F.	= Bottom Flat	M. D.	= Mean Diameter
T. F.	= Top Flat	Flg.	= Flange
Sym.	= Symmetrical	Std.	= Standard
E. T.	= Equal Taper	Cu.	= Cubic
C. T.	= Center Taper	Sq.	= Square
ID	= Inside Diameter	Rd.	= Round
I. D.	= Inside Dimension	Rect.	= Rectangle
OD	= Outside Diameter	Sup.	= Supply
O. D.	= Outside Dimension	R. A.	= Return Air
O. A.	= Overall Area	Cir.	= Circular
S. O.	= Stretchout	L.	= Length
Dn.	= Down	Off. or Set.	= Offset
Mk.	= Mark	Pcs.	= Pieces
S. V.	= Single Vanes	Req.	= Required
D. V.	= Double Vanes	Lbs.	= Pounds
Sq. Ft.	= Square Feet	Galv.	= Galvanized
Sq. In.	= Square Inches	Elev.	= Elevation
Circ.	= Circumference	Pitts.	= Pittsburg
Dia.	= Diameter	S. E.	= Single Edge
R. or Rad.	= Radius	D. S.	= Drive Slip
Plstr. Gd.	= Plaster Ground	Pat.	= Pattern
Deg.	= Degree	Seg.	= Segment
Ga.	= Gauge	Cap.	= Capacity

chapter XIV

REFERENCE TABLES

DECIMAL EQUIVALENTS						
		¼₄	.0156		³³⁄₆₄	.5156
	¹⁄₃₂		.0312	¹⁷⁄₃₂		.5312
		³⁄₆₄	.0468		³⁵⁄₆₄	.5468
¹⁄₁₆			.0625	⁹⁄₁₆		.5625
		⁵⁄₆₄	.0781		³⁷⁄₆₄	.5781
	³⁄₃₂		.0937	¹⁹⁄₃₂		.5937
		⁷⁄₆₄	.1093		³⁹⁄₆₄	.6093
⅛			.125	⅝		.625
		⁹⁄₆₄	.1406		⁴¹⁄₆₄	.6406
	⁵⁄₃₂		.1562	²¹⁄₃₂		.6562
		¹¹⁄₆₄	.1718		⁴³⁄₆₄	.6718
³⁄₁₆			.1875	¹¹⁄₁₆		.6875
		¹³⁄₆₄	.2031		⁴⁵⁄₆₄	.7031
	⁷⁄₃₂		.2187	²³⁄₃₂		.7187
		¹⁵⁄₆₄	.2343		⁴⁷⁄₆₄	.7343
¼			.250	¾		.750
		¹⁷⁄₆₄	.2656		⁴⁹⁄₆₄	7656
	⁹⁄₃₂		.2812	²⁵⁄₃₂		.7812
		¹⁹⁄₆₄	.2968		⁵¹⁄₆₄	.7968
⁵⁄₁₆			.3125	¹³⁄₁₆		.8125
		²¹⁄₆₄	.3281		⁵³⁄₆₄	.8281
	¹¹⁄₃₂		.3437	²⁷⁄₃₂		.8437
		²³⁄₆₄	.3593		⁵⁵⁄₆₄	.8593
⅜			.375	⅞		.875
		²⁵⁄₆₄	.3906		⁵⁷⁄₆₄	.8906
	¹³⁄₃₂		.4062	²⁹⁄₃₂		.9062
		²⁷⁄₆₄	.4218		⁵⁹⁄₆₄	.9218
⁷⁄₁₆			.4375	¹⁵⁄₁₆		.9375
		²⁹⁄₆₄	.4531		⁶¹⁄₆₄	.9531
	¹⁵⁄₃₂		.4687	³¹⁄₃₂		.9687
		³¹⁄₆₄	.4843		⁶³⁄₆₄	.9843
½			.500			

Reference Tables

STRETCHOUTS
FOR THROAT AND HEEL WRAPPERS

RADIUS	DEGREE OF ANGLE			
	90°	60°	45°	30°
1	1⁹⁄₁₆	1¹⁄₁₆	¹³⁄₁₆	½
2	3⅛	2⅛	1⁹⁄₁₆	1¹⁄₁₆
3	4¾	3⅛	2⁹⁄₁₆	1⁹⁄₁₆
4	6¼	4³⁄₁₆	3⅛	2⅛
5	7⅞	5¼	3¹⁵⁄₁₆	2⅝
6	9⁷⁄₁₆	6⁵⁄₁₆	4¹¹⁄₁₆	3⅛
7	11	7⁵⁄₁₆	5½	3¹¹⁄₁₆
8	12⁹⁄₁₆	8⅜	6⁵⁄₁₆	4³⁄₁₆
9	14⅛	9⁷⁄₁₆	7¹⁄₁₆	4¹¹⁄₁₆
10	15¹¹⁄₁₆	10½	7⅞	5¼
11	17¼	11½	8⅝	5¾
12	18⅞	12⁹⁄₁₆	9⁷⁄₁₆	6⁵⁄₁₆
13	20⁷⁄₁₆	13⅝	10³⁄₁₆	6¹³⁄₁₆
14	22	14¹¹⁄₁₆	11	7⁵⁄₁₆
15	23⁹⁄₁₆	15¹¹⁄₁₆	11¾	7⅞
16	25⅛	16¾	12⁹⁄₁₆	8⅜
17	26¹¹⁄₁₆	17¹³⁄₁₆	13⅜	8⅞
18	28¼	18⅞	14⅛	9⁷⁄₁₆
19	29⅞	19⅞	14¹⁵⁄₁₆	9¹⁵⁄₁₆
20	31⁷⁄₁₆	20¹⁵⁄₁₆	15¹¹⁄₁₆	10½
21	33	22	16½	11
22	34⁹⁄₁₆	23¹⁄₁₆	17¼	11½
23	36⅛	24¹⁄₁₆	18¹⁄₁₆	12¹⁄₁₆
24	37¹¹⁄₁₆	25⅛	18⅞	12⁹⁄₁₆
25	39¼	26³⁄₁₆	19⅝	13¹⁄₁₆
26	40⅞	27¼	20⁷⁄₁₆	13⅝
27	42⁷⁄₁₆	28¼	21³⁄₁₆	14⅛
28	44	29⁵⁄₁₆	22	14¹¹⁄₁₆
29	45⁹⁄₁₆	30⅜	22¾	15³⁄₁₆
30	47⅛	31⁷⁄₁₆	23⁹⁄₁₆	15¹¹⁄₁₆
31	48¹¹⁄₁₆	32⁷⁄₁₆	24⅜	16¼
32	50¼	33½	25⅛	16¾

RADIUS	90°	60°	45°	30°
33	51 13/16	34 9/16	25 15/16	17 1/4
34	53 3/8	35 5/8	26 11/16	17 13/16
35	54 15/16	36 5/8	27 1/2	18 5/16
36	56 9/16	37 11/16	28 1/4	18 7/8
37	58 1/8	38 3/4	29 1/16	19 3/8
38	59 11/16	39 13/16	29 7/8	19 7/8
39	61 1/4	40 13/16	30 5/8	20 7/16
40	62 3/4	41 7/8	31 7/16	20 15/16
41	64 3/8	43 15/16	32 15/16	21 7/16
42	65 15/16	44	33	22
43	67 1/2	45	33 3/4	22 1/2
44	69 1/8	46 1/16	34 9/16	23 1/16
45	70 11/16	47 1/8	35 5/16	23 9/16
46	72 1/4	48 3/16	36 1/8	24 1/16
47	73 13/16	49 3/16	36 15/16	24 5/8
48	75 3/8	50 1/4	37 11/16	25 1/8
49	76 15/16	51 5/16	38 1/2	25 5/8
50	78 1/2	52 3/8	39 1/4	26 3/16
51	80 1/16	53 7/16	40 1/16	26 11/16
52	81 11/16	54 7/16	40 13/16	27 1/4
53	83 1/4	55 1/2	41 5/8	27 3/4
54	84 13/16	56 9/16	42 7/16	28 1/4
55	86 3/8	57 5/8	43 3/16	28 13/16
56	87 15/16	58 5/8	44	29 5/16
57	89 1/2	59 11/16	44 3/4	29 7/8
58	91 1/16	60 3/4	45 9/16	30 3/8
59	92 11/16	61 13/16	46 5/16	30 7/8
60	94 1/4	62 13/16	47 1/8	31 7/16
61	95 13/16	63 7/8	47 15/16	31 15/16
62	97 3/8	64 15/16	48 11/16	32 7/16
63	98 15/16	66	49 1/2	33
64	100 1/2	67	50 1/4	33 1/2
65	102 1/16	68 1/16	51 1/16	34 1/16
66	103 5/8	69 1/8	51 13/16	34 9/16
67	105 3/16	70 3/16	52 5/8	35 1/16
68	106 3/4	71 3/16	53 7/16	35 5/8
69	108 3/8	72 1/4	54 3/16	36 1/8
70	109 15/16	73 5/16	55	36 11/16
71	111 1/2	74 3/8	55 3/4	37 3/16
72	113 1/16	75 3/8	56 9/16	37 11/16

Reference Tables

RADIUS	90°	60°	45°	30°
73	114$\frac{5}{8}$	76$\frac{7}{16}$	57$\frac{5}{16}$	38$\frac{1}{4}$
74	116$\frac{3}{16}$	77$\frac{1}{2}$	58$\frac{1}{8}$	38$\frac{3}{4}$
75	117$\frac{3}{4}$	78$\frac{9}{16}$	58$\frac{7}{8}$	39$\frac{1}{4}$
76	119$\frac{3}{8}$	79$\frac{9}{16}$	59$\frac{11}{16}$	39$\frac{3}{16}$
77	120$\frac{15}{16}$	80$\frac{5}{8}$	60$\frac{1}{2}$	40$\frac{5}{16}$
78	122$\frac{1}{2}$	81$\frac{11}{16}$	61$\frac{1}{4}$	40$\frac{13}{16}$
79	124$\frac{1}{16}$	82$\frac{3}{4}$	62$\frac{1}{16}$	41$\frac{3}{8}$
80	125$\frac{5}{8}$	83$\frac{3}{4}$	62$\frac{13}{16}$	41$\frac{7}{8}$
81	127$\frac{3}{16}$	84$\frac{13}{16}$	63$\frac{5}{8}$	42$\frac{7}{16}$
82	128$\frac{3}{4}$	85$\frac{7}{8}$	64$\frac{3}{8}$	42$\frac{15}{16}$
83	130$\frac{3}{8}$	86$\frac{15}{16}$	65$\frac{3}{16}$	43$\frac{7}{16}$
84	131$\frac{7}{8}$	87$\frac{15}{16}$	66	44
85	133$\frac{1}{2}$	89	66$\frac{3}{4}$	44$\frac{1}{2}$
86	135$\frac{1}{16}$	90$\frac{1}{16}$	67$\frac{9}{16}$	45
87	136$\frac{5}{8}$	91$\frac{1}{8}$	68$\frac{5}{16}$	45$\frac{5}{16}$
88	138$\frac{1}{4}$	92$\frac{1}{8}$	69$\frac{1}{8}$	46$\frac{1}{16}$
89	139$\frac{13}{16}$	93$\frac{3}{16}$	69$\frac{7}{8}$	46$\frac{5}{8}$
90	141$\frac{3}{8}$	94$\frac{1}{4}$	70$\frac{11}{16}$	47$\frac{1}{8}$
91	142$\frac{15}{16}$	95$\frac{5}{16}$	71$\frac{1}{2}$	47$\frac{5}{8}$
92	144$\frac{1}{2}$	96$\frac{5}{16}$	72$\frac{1}{4}$	48$\frac{3}{16}$
93	146$\frac{1}{16}$	97$\frac{3}{8}$	73$\frac{1}{16}$	48$\frac{11}{16}$
94	147$\frac{11}{16}$	98$\frac{7}{16}$	73$\frac{13}{16}$	49$\frac{3}{16}$
95	149$\frac{1}{4}$	99$\frac{1}{2}$	74$\frac{5}{8}$	49$\frac{3}{4}$
96	150$\frac{13}{16}$	100$\frac{1}{2}$	75$\frac{3}{8}$	50$\frac{1}{4}$
97	152$\frac{3}{8}$	101$\frac{9}{16}$	76$\frac{3}{16}$	50$\frac{13}{16}$
98	153$\frac{15}{16}$	102$\frac{5}{8}$	77	51$\frac{5}{16}$
99	155$\frac{1}{2}$	103$\frac{11}{16}$	77$\frac{3}{4}$	51$\frac{13}{16}$
100	157$\frac{1}{16}$	104$\frac{11}{16}$	78$\frac{9}{16}$	52$\frac{3}{8}$

STRETCHOUT $= .01745$ x Radius x Degrees

NOTE: *These figures are given in the nearest workable fraction, making it easier to apply in actual shop practice. Allowance for lock and flange MUST BE ADDED to these figures as required before blocking out material.*

DUCT EQUIVALENTS

ROUND EQUIVALENTS OF RECTANGULAR DUCT (Equal Friction)

DUCT SIDES	5	6	7	8	9	10	11	12	13	14	15	16	17	18	19	20	21	22	23	24
8	6.9	7.6	8.2	8.8																
9	7.3	8.0	8.7	9.3	9.9															
10	7.7	8.4	9.2	9.8	10.4	11.0														
11	8.0	8.8	9.6	10.2	10.9	11.5	12.1													
12	8.3	9.2	10.0	10.7	11.4	12.0	12.6	13.2												
13	8.7	9.6	10.4	11.1	11.8	12.5	13.1	13.7	14.3											
14	8.9	9.9	10.8	11.5	12.3	12.9	13.6	14.3	14.9	15.4										
15	9.2	10.2	11.1	11.9	12.7	13.4	14.1	14.7	15.3	16.0	16.5									
16	9.5	10.5	11.4	12.3	13.1	13.8	14.5	15.2	15.8	16.5	17.1	17.6								
17	9.8	10.8	11.8	12.6	13.5	14.2	15.0	15.7	16.3	17.0	17.6	18.2	18.7							
18	10.0	11.1	12.1	13.0	13.8	14.6	15.4	16.1	16.8	17.4	18.1	18.7	19.2	19.8						
19	10.3	11.4	12.4	13.3	14.2	15.0	15.8	16.5	17.2	17.9	18.6	19.2	19.8	20.4	20.9					
20	10.5	11.6	12.7	13.6	14.5	15.4	16.2	17.0	17.6	18.4	19.0	19.7	20.3	20.9	21.5	22.0				
22	11.0	12.1	13.2	14.2	15.2	16.1	16.9	17.8	18.5	19.2	19.9	20.6	21.3	21.9	22.5	23.1	23.6	24.2		
24	11.4	12.6	13.8	14.8	15.8	16.8	17.6	18.5	19.3	20.0	20.8	21.5	22.2	22.8	23.5	24.0	24.7	25.2	25.9	26.4
26	11.8	13.1	14.3	15.4	16.4	17.3	18.3	19.2	20.0	20.8	21.6	22.3	23.0	23.8	24.4	25.1	25.7	26.3	26.9	27.5
28	12.2	13.5	14.8	15.9	17.0	18.0	19.0	19.8	20.7	21.5	22.4	23.1	23.9	24.6	25.3	26.0	26.6	27.3	27.9	28.5
30	12.6	13.9	15.2	16.4	17.5	18.5	19.5	20.5	21.4	22.2	23.1	23.9	24.7	25.4	26.2	26.8	27.5	28.2	28.8	29.5
32	12.9	14.3	15.6	16.9	18.0	19.1	20.1	21.1	22.0	22.9	23.8	24.6	25.4	26.2	27.0	27.7	28.4	29.1	29.8	30.5
34	13.2	14.7	16.1	17.3	18.5	19.6	20.7	21.6	22.6	23.5	24.4	25.3	26.2	26.9	27.7	28.5	29.2	30.0	30.7	31.3
36	13.6	15.1	16.4	17.7	19.0	20.1	21.2	22.2	23.2	24.2	25.1	26.0	26.8	27.7	28.5	29.3	30.0	30.8	31.5	32.2
38	13.9	15.4	16.8	18.2	19.4	20.6	21.7	22.7	23.8	24.8	25.8	26.7	27.5	28.4	29.2	30.0	30.8	31.5	32.3	33.1
40	14.3	15.7	17.2	18.6	19.8	21.1	22.2	23.3	24.4	25.4	26.4	27.3	28.2	29.1	29.9	30.8	31.6	32.4	33.1	33.9
42	14.5	16.1	17.6	19.0	20.3	21.6	22.7	23.8	24.9	25.9	26.9	27.9	28.8	29.8	30.7	31.4	32.2	33.0	33.8	34.5
44	14.8	16.4	18.0	19.4	20.7	22.0	23.1	24.3	25.4	26.5	27.5	28.5	29.5	30.3	31.2	32.1	32.9	33.7	34.5	35.3
46	15.1	16.7	18.4	19.8	21.1	22.4	23.6	24.8	25.9	27.0	28.1	29.1	30.1	31.0	31.9	32.8	33.8	34.6	35.4	36.2
48	15.4	17.0	18.7	20.1	21.5	22.8	24.1	25.2	26.4	27.5	28.6	29.6	30.5	31.6	32.5	33.4	34.3	35.2	36.1	37.0
50	15.7	17.3	19.0	20.4	21.9	23.2	24.5	25.7	26.9	28.0	29.2	30.3	31.3	32.2	33.1	34.1	35.0	35.9	36.8	37.6
52	15.9	17.6	19.2	20.8	22.2	23.6	24.9	26.2	27.4	28.5	29.6	30.7	31.8	32.9	33.8	34.7	35.6	36.5	37.4	38.3
54	16.1	17.9	19.6	21.1	22.6	24.0	25.3	26.6	27.8	29.0	30.1	31.2	32.3	33.4	34.4	35.3	36.3	37.2	38.0	38.9
56	16.3	18.2	19.9	21.5	22.9	24.4	25.7	27.0	28.3	29.5	30.6	31.7	32.8	33.9	34.9	35.9	36.9	37.8	38.7	39.6
58	16.6	18.4	20.2	21.8	23.3	24.7	26.1	27.4	28.7	30.0	31.1	32.2	33.3	34.4	35.4	36.4	37.4	38.4	39.4	40.3

Reference Tables

SIDES	NET AREAS OF DUCT												
	12″	14″	16″	18″	20″	22″	24″	26″	28″	30″	32″	34″	36″
5″	.42	.49	.55	.62	.69	.76	.83	.90	.97	1.04	1.11	1.18	1.25
6″	.50	.58	.67	.75	.83	.92	1.00	1.08	1.17	1.25	1.33	1.42	1.50
7″	.58	.68	.78	.87	.97	1.07	1.17	1.26	1.36	1.46	1.55	1.65	1.75
8″	.67	.78	.89	1.00	1.11	1.22	1.33	1.44	1.55	1.67	1.78	1.89	2.00
9″	.75	.88	1.00	1.12	1.25	1.37	1.50	1.62	1.75	1.87	2.00	2.12	2.25
10″	.83	.97	1.11	1.25	1.39	1.52	1.66	1.80	1.94	2.08	2.22	2.36	2.50
11″	.92	1.07	1.22	1.37	1.53	1.68	1.83	1.99	2.14	2.29	2.44	2.60	2.75
12″	1.00	1.16	1.33	1.50	1.67	1.83	2.00	2.16	2.33	2.50	2.67	2.83	3.00
13″	1.08	1.26	1.44	1.62	1.80	1.98	2.16	2.34	2.53	2.71	2.88	3.07	3.25
14″	1.16	1.36	1.55	1.75	1.94	2.14	2.33	2.53	2.72	2.92	3.11	3.30	3.50
15″	1.25	1.46	1.67	1.87	2.08	2.29	2.50	2.71	2.92	3.12	3.34	3.54	3.74
16″	1.33	1.55	1.78	2.00	2.22	2.44	2.66	2.89	3.11	3.33	3.55	3.78	4.00
18″	1.50	1.75	2.00	2.25	2.50	2.75	3.00	3.25	3.50	3.75	4.00	4.25	4.50
20″	1.67	1.94	2.22	2.50	2.78	3.06	3.33	3.61	3.89	4.17	4.44	4.72	5.00
22″	1.83	2.14	2.44	2.75	3.05	3.36	3.67	3.97	4.27	4.58	4.88	5.19	5.50
24″	2.00	2.33	2.66	3.00	3.33	3.67	4.00	4.33	4.67	5.00	5.33	5.67	6.00
26″	2.17	2.53	2.89	3.25	3.61	3.97	4.33	4.69	5.06	5.42	5.77	6.14	6.50
28″	2.33	2.72	3.11	3.50	3.89	4.27	4.67	5.06	5.45	5.84	6.23	6.61	7.00
30″	2.50	2.92	3.33	3.75	4.17	4.58	5.00	5.42	5.84	6.25	6.67	7.08	7.50
32″	2.67	3.11	3.55	4.00	4.44	4.88	5.33	5.77	6.23	6.67	7.12	7.56	8.00
34″	2.83	3.30	3.78	4.25	4.72	5.19	5.67	6.14	6.61	7.08	7.56	8.02	8.50
36″	3.00	3.50	4.00	4.50	5.00	5.50	6.00	6.50	7.00	7.50	8.00	8.50	9.00
38″	3.16	3.70	4.22	4.75	5.27	5.80	6.33	6.86	7.39	7.92	8.44	8.97	9.50
40″	3.33	3.89	4.44	5.00	5.55	6.10	6.67	7.22	7.78	8.33	8.88	9.45	10.00
42″	3.50	4.08	4.66	5.25	5.83	6.41	7.00	7.58	8.17	8.75	9.33	9.92	10.50
44″	3.66	4.27	4.89	5.50	6.11	6.72	7.33	7.95	8.56	9.17	9.77	10.40	11.00
46″	3.83	4.47	5.11	5.75	6.39	7.02	7.67	8.30	8.95	9.58	10.22	10.85	11.50
48″	4.00	4.67	5.33	6.00	6.66	7.32	8.00	8.66	9.34	10.00	10.67	11.35	12.00
50″	4.16	4.86	5.55	6.25	6.94	7.64	8.33	9.03	9.73	10.42	11.11	11.80	12.50
52″	4.33	5.05	5.78	6.50	7.22	7.95	8.67	9.30	10.11	10.82	11.55	12.30	13.00
54″	4.50	5.25	6.00	6.75	7.50	8.25	9.00	9.75	10.50	11.24	12.00	12.75	13.50
56″	4.66	5.45	6.22	7.00	7.78	8.55	9.33	10.09	10.89	11.66	12.44	13.22	14.00
58″	4.83	5.64	6.44	7.25	8.05	8.86	9.67	10.49	11.25	12.07	12.68	13.70	14.50
60″	5.00	5.84	6.66	7.50	8.33	9.16	10.00	10.83	11.66	12.50	13.35	14.15	15.00
62″	5.16	6.02	6.89	7.75	8.61	9.47	10.34	11.19	12.05	12.92	13.75	14.65	15.50
64″	5.33	6.22	7.11	8.00	8.89	9.77	10.66	11.55	12.45	13.34	14.21	15.12	16.00
66″	5.50	6.41	7.33	8.25	9.17	10.09	11.00	11.92	12.83	13.76	14.67	15.59	16.50
68″	5.66	6.60	7.55	8.50	9.44	10.40	11.30	12.30	13.25	14.16	15.12	16.05	17.00
70″	5.83	6.80	7.77	8.73	9.72	10.70	11.68	12.66	13.62	14.58	15.55	16.54	17.50
72″	6.00	7.00	8.00	9.00	10.00	11.00	12.00	13.00	14.00	15.00	16.00	17.00	18.00
74″	6.16	7.20	8.22	9.25	10.28	11.31	12.34	13.36	14.39	15.40	16.45	17.49	18.49
76″	6.33	7.38	8.44	9.50	10.55	11.60	12.65	13.70	14.79	14.85	16.85	17.92	19.00
78″	6.50	7.57	8.66	9.75	10.81	11.91	13.00	14.08	15.18	16.25	17.35	18.42	19.50
80″	6.66	7.77	8.87	10.00	11.10	12.20	13.30	14.40	15.50	16.70	17.80	18.86	20.00

SIDES	NET AREAS OF DUCT											
	38"	40"	42"	44"	46"	48"	50"	52"	54"	56"	58"	60"
5"	1.32	1.39	1.46	1.53	1.60	1.67	1.74	1.81	1.88	1.95	2.01	2.08
6"	1.58	1.67	1.75	1.83	1.92	2.00	2.08	2.17	2.25	2.34	2.42	2.50
7"	1.85	1.94	2.04	2.14	2.24	2.33	2.43	2.53	2.62	2.72	2.82	2.91
8"	2.11	2.22	2.33	2.44	2.55	2.67	2.78	2.89	3.00	3.11	3.22	3.33
9"	2.37	2.50	2.62	2.75	2.87	3.00	3.12	3.25	3.37	3.50	3.62	3.75
10"	2.64	2.78	2.91	3.05	3.19	3.33	3.47	3.61	3.75	3.89	4.03	4.17
11"	2.90	3.06	3.21	3.36	3.51	3.67	3.82	3.97	4.12	4.28	4.43	4.58
12"	3.16	3.33	3.50	3.67	3.83	4.00	4.16	4.33	4.50	4.67	4.83	5.00
13"	3.43	3.61	3.79	3.97	4.15	4.33	4.51	4.69	4.87	5.05	5.24	5.42
14"	3.70	3.89	4.08	4.27	4.47	4.67	4.86	5.05	5.25	5.45	5.64	6.84
15"	3.96	4.17	4.37	4.58	4.80	5.00	5.21	5.41	5.62	5.83	6.04	6.25
16"	4.22	4.44	4.66	4.89	5.11	5.33	5.55	5.78	6.00	6.22	6.44	6.66
18"	4.75	5.00	5.25	5.50	5.75	6.00	6.25	6.50	6.75	7.00	7.25	7.50
20"	5.27	5.55	5.83	6.11	6.39	6.66	6.94	7.22	7.50	7.78	8.05	8.33
22"	5.80	6.10	6.41	6.72	7.02	7.32	7.64	7.95	8.25	8.55	8.86	9.16
24"	6.33	6.67	7.00	7.33	7.67	8.00	8.33	8.67	9.00	9.33	9.67	10.00
26"	6.86	7.22	7.58	7.95	8.30	8.66	9.03	9.38	9.75	10.09	10.49	10.83
28"	7.39	7.78	8.17	8.56	8.95	9.34	9.73	10.11	10.50	10.89	11.25	11.66
30"	7.92	8.33	8.75	9.17	9.58	10.00	10.42	10.82	11.24	11.66	12.07	12.50
32"	8.44	8.88	9.33	9.77	10.22	10.67	11.11	11.55	12.00	12.44	12.88	13.35
34"	8.97	9.45	9.92	10.40	10.85	11.35	11.80	12.30	12.75	13.22	12.70	14.15
36"	9.50	10.00	10.50	11.00	11.50	12.00	12.50	13.00	13.50	14.00	14.50	15.00
38"	10.02	10.55	11.08	11.62	12.15	12.65	13.20	13.72	14.25	14.80	15.30	15.80
40"	10.55	11.10	11.66	12.20	12.79	13.34	13.89	14.44	15.00	15.65	16.10	16.66
42"	11.08	11.66	12.25	12.83	13.44	14.00	14.58	15.15	15.75	16.31	16.93	17.50
44"	11.62	12.20	12.83	13.45	14.05	14.68	15.28	15.88	16.50	17.11	17.72	18.35
46"	12.15	12.79	13.44	14.05	14.70	15.34	16.00	16.61	17.25	17.90	18.52	19.15
48"	12.65	13.34	14.00	14.68	15.34	16.00	16.66	17.31	18.00	18.66	19.35	20.00
50"	13.20	13.89	14.58	15.28	16.00	16.66	17.35	18.05	18.75	19.43	20.03	20.80
52"	13.72	14.44	15.15	15.88	16.61	17.31	18.05	18.75	19.50	20.20	20.95	21.68
54"	14.25	15.00	15.75	16.50	17.25	18.00	18.75	19.50	20.25	21.00	21.79	22.50
56"	14.80	15.55	16.31	17.11	19.70	18.66	19.43	20.20	21.00	21.80	22.52	23.29
58"	15.30	16.10	16.93	17.72	18.52	19.35	20.09	20.95	21.79	22.52	23.38	24.11
60"	15.80	16.66	17.50	18.35	19.15	20.00	20.80	21.65	22.50	23.29	24.11	25.00
62"	16.36	17.21	18.10	18.95	19.82	20.65	21.50	22.40	23.24	24.10	25.00	25.80
64"	16.86	17.80	18.66	19.57	20.45	21.31	22.22	23.12	24.00	24.86	25.80	26.68
66"	17.40	18.35	19.25	20.19	21.08	22.00	22.92	23.85	24.70	25.65	26.60	27.50
68"	17.95	18.87	19.80	20.80	21.70	22.65	23.60	24.56	25.50	26.43	27.40	28.32
70"	18.45	19.43	20.40	21.38	22.32	23.31	24.30	25.30	26.22	27.21	28.20	29.17
72"	19.00	20.00	21.00	22.00	23.00	24.00	25.00	26.00	27.00	28.00	29.00	30.00
74"	19.55	20.58	21.59	22.60	23.62	24.65	25.70	26.71	27.77	28.79	29.81	30.81
76"	20.05	21.10	22.19	23.20	24.30	25.30	26.40	27.41	28.45	29.59	30.61	31.65
78"	20.60	21.63	22.72	23.80	24.90	26.00	27.03	28.13	29.28	30.39	31.40	32.50
80"	21.10	22.21	23.30	24.40	25.60	26.70	27.80	28.90	30.00	31.10	32.20	33.30

Reference Tables

CIRCUMFERENCES AND AREAS OF CIRCLES

Diam. In.	Circum. In.	Area Sq. In.	Diam. In.	Circum. In.	Area Sq. In.	Diam. In.	Circum. In.	Area Sq. In.	Diam. In.	Circum. In.	Area Sq. In.
1	3.141	.7	5 1/2	17.279	23.7	14	43.982	153.9	23	72.257	415.4
1 1/16	3.337	.8	5 9/16	17.475	24.3	14 1/8	44.375	156.7	23 1/8	72.649	420.0
1 1/8	3.534	.9	5 5/8	17.671	24.8	14 1/4	44.768	159.4	23 1/4	73.042	424.5
1 3/16	3.730	1.1	5 11/16	17.867	25.4	14 3/8	45.160	162.3	23 3/8	73.435	429.1
1 1/4	3.927	1.2	5 3/4	18.064	25.9	14 1/2	45.553	165.1	23 1/2	73.827	433.7
1 5/16	4.123	1.3	5 13/16	18.261	26.5	14 5/8	45.946	167.9	23 5/8	74.220	438.3
1 3/8	4.319	1.4	5 7/8	18.457	27.1	14 3/4	46.338	170.8	23 3/4	74.613	443.0
1 7/16	4.516	1.6	5 15/16	18.653	27.6	14 7/8	46.731	173.7	23 7/8	75.006	447.6
1 1/2	4.712	1.7	6	18.850	28.2	15	47.124	176.7	24	75.398	452.3
1 9/16	4.908	1.9	6 1/8	19.242	29.4	15 1/8	47.517	179.6	24 1/8	75.791	457.1
1 5/8	5.105	2.0	6 1/4	19.635	30.6	15 1/4	47.909	182.6	24 1/4	76.184	461.8
1 11/16	5.301	2.2	6 3/8	20.028	31.9	15 3/8	48.302	185.6	24 3/8	76.576	466.6
1 3/4	5.497	2.4	6 1/2	20.420	33.1	15 1/2	48.695	188.6	24 1/2	76.969	471.4
1 13/16	5.694	2.5	6 5/8	20.813	34.4	15 5/8	49.087	191.7	24 5/8	77.362	476.2
1 7/8	5.890	2.7	6 3/4	21.206	35.7	15 3/4	49.480	194.8	24 3/4	77.754	481.1
1 15/16	6.086	2.9	6 7/8	21.598	37.1	15 7/8	49.873	197.9	24 7/8	78.147	485.9
2	6.283	3.1	7	21.991	38.4	16	50.265	201.0	25	78.540	490.8
2 1/16	6.479	3.3	7 1/8	22.384	39.8	16 1/8	50.658	204.2	25 1/8	78.933	495.7
2 1/8	6.675	3.5	7 1/4	22.776	41.2	16 1/4	51.051	207.3	25 1/4	79.325	500.7
2 3/16	6.872	3.7	7 3/8	23.169	42.7	16 3/8	51.444	210.6	25 3/8	79.718	505.7
2 1/4	7.068	3.9	7 1/2	23.562	44.1	16 1/2	51.836	213.8	25 1/2	80.111	510.7
2 5/16	7.264	4.2	7 5/8	23.955	45.6	16 5/8	52.229	217.0	25 5/8	80.503	515.7
2 3/8	7.461	4.4	7 3/4	24.347	47.1	16 3/4	52.622	220.3	25 3/4	80.896	520.7
2 7/16	7.657	4.6	7 7/8	24.740	48.7	16 7/8	53.014	223.6	25 7/8	81.289	525.8
2 1/2	7.854	4.9	8	25.133	50.2	17	53.407	226.9	26	81.681	530.9
2 9/16	8.050	5.1	8 1/8	25.525	51.8	17 1/8	53.800	230.3	26 1/8	82.074	536.0
2 5/8	8.246	5.4	8 1/4	25.918	53.4	17 1/4	54.192	233.7	26 1/4	82.467	541.1
2 11/16	8.443	5.6	8 3/8	26.311	55.0	17 3/8	54.585	237.1	26 3/8	82.860	546.3
2 3/4	8.639	5.9	8 1/2	26.704	56.7	17 1/2	54.978	240.5	26 1/2	83.252	551.5
2 13/16	8.835	6.2	8 5/8	27.096	58.4	17 5/8	55.371	243.9	26 5/8	83.645	556.7
2 7/8	9.032	6.4	8 3/4	27.489	60.1	17 3/4	55.763	247.4	26 3/4	84.038	562.0
2 15/16	9.228	6.7	8 7/8	27.882	61.8	17 7/8	56.156	250.9	26 7/8	84.430	567.2
3	9.424	7.0	9	28.274	63.6	18	56.549	254.4	27	84.823	572.5
3 1/16	9.621	7.3	9 1/8	28.667	65.3	18 1/8	56.941	258.0	27 1/8	85.216	577.8
3 1/8	9.817	7.6	9 1/4	29.060	67.2	18 1/4	57.334	261.5	27 1/4	85.608	583.2
3 3/16	10.014	7.9	9 3/8	29.452	69.0	18 3/8	57.727	265.1	27 3/8	86.001	588.5
3 1/4	10.210	8.2	9 1/2	29.845	70.8	18 1/2	58.119	268.8	27 1/2	86.394	593.9
3 5/16	10.407	8.6	9 5/8	30.238	72.7	18 5/8	58.512	272.4	27 5/8	86.786	599.3
3 3/8	10.603	8.9	9 3/4	30.631	74.6	18 3/4	58.905	276.1	27 3/4	87.179	604.8
3 7/16	10.799	9.2	9 7/8	31.023	76.5	18 7/8	59.298	279.8	27 7/8	87.572	610.2
3 1/2	10.996	9.6	10	31.416	78.5	19	59.690	283.5	28	87.965	615.7
3 9/16	11.192	9.9	10 1/8	31.809	80.5	19 1/8	60.083	287.2	28 1/8	88.357	621.2
3 5/8	11.388	10.3	10 1/4	32.201	82.5	19 1/4	60.476	291.0	28 1/4	88.750	626.8
3 11/16	11.585	10.6	10 3/8	32.594	84.5	19 3/8	60.868	294.8	28 3/8	89.143	632.3
3 3/4	11.781	11.0	10 1/2	32.987	86.5	19 1/2	61.261	298.6	28 1/2	89.535	637.9
3 13/16	11.977	11.4	10 5/8	33.379	88.6	19 5/8	61.654	302.4	28 5/8	89.928	643.5
3 7/8	12.174	11.7	10 3/4	33.772	90.7	19 3/4	62.046	306.3	28 3/4	90.321	649.1
3 15/16	12.370	12.1	10 7/8	34.165	92.8	19 7/8	62.439	310.2	28 7/8	90.713	654.8
4	12.566	12.5	11	34.558	95.0	20	62.832	314.1	29	91.106	660.5
4 1/16	12.763	12.9	11 1/8	34.950	97.2	20 1/8	63.225	318.1	29 1/8	91.499	666.2
4 1/8	12.959	13.3	11 1/4	35.343	99.4	20 1/4	63.617	322.0	29 1/4	91.892	671.9
4 3/16	13.155	13.7	11 3/8	35.736	101.6	20 3/8	64.010	326.0	29 3/8	92.284	677.7
4 1/4	13.352	14.1	11 1/2	36.128	103.8	20 1/2	64.403	330.0	29 1/2	92.677	683.4
4 5/16	13.548	14.6	11 5/8	36.521	106.1	20 5/8	64.795	334.1	29 5/8	93.070	689.3
4 3/8	13.744	15.0	11 3/4	36.914	108.4	20 3/4	65.188	338.1	29 3/4	93.462	695.1
4 7/16	13.941	15.4	11 7/8	37.306	110.7	20 7/8	65.581	342.2	29 7/8	93.855	700.9
4 1/2	14.137	15.9	12	37.699	113.1	21	65.973	346.3	30	94.248	706.8
4 9/16	14.334	16.3	12 1/8	38.092	115.4	21 1/8	66.366	350.5	30 1/8	94.640	712.7
4 5/8	14.530	16.8	12 1/4	38.485	117.8	21 1/4	66.759	354.6	30 1/4	95.033	718.6
4 11/16	14.726	17.2	12 3/8	38.877	120.2	21 3/8	67.152	358.8	30 3/8	95.426	724.6
4 3/4	14.923	17.7	12 1/2	39.270	122.7	21 1/2	67.544	363.0	30 1/2	95.819	730.6
4 13/16	15.119	18.1	12 5/8	39.663	125.1	21 5/8	67.937	367.2	30 5/8	96.211	736.6
4 7/8	15.315	18.6	12 3/4	40.055	127.6	21 3/4	68.330	371.5	30 3/4	96.604	742.6
4 15/16	15.512	19.1	12 7/8	40.448	130.1	21 7/8	68.722	375.8	30 7/8	96.997	748.6
5	15.708	19.6	13	40.841	132.7	22	69.115	380.1	31	97.389	754.7
5 1/16	15.904	20.1	13 1/8	41.233	135.3	22 1/8	69.508	384.4	31 1/8	97.782	760.8
5 1/8	16.101	20.6	13 1/4	41.626	137.8	22 1/4	69.900	388.8	31 1/4	98.175	766.9
5 3/16	16.297	21.1	13 3/8	42.019	140.5	22 3/8	70.293	393.2	31 3/8	98.567	773.1
5 1/4	16.493	21.6	13 1/2	42.412	143.1	22 1/2	70.686	397.6	31 1/2	98.960	779.3
5 5/16	16.690	22.1	13 5/8	42.804	145.8	22 5/8	71.079	402.0	31 5/8	99.353	785.5
5 3/8	16.886	22.6	13 3/4	43.197	148.4	22 3/4	71.471	406.4	31 3/4	99.746	791.7
5 7/16	17.082	23.2	13 7/8	43.590	151.2	22 7/8	71.864	410.9	31 7/8	100.138	797.9

Reference Tables

CIRCUMFERENCES AND AREAS OF CIRCLES

Diam. In.	Circum. In.	Area Sq. In.	Diam. In.	Circum. In.	Area Sq. In.	Diam. In.	Circum. In.	Area Sq. In.	Diam. In.	Circum. In.	Area Sq. In.	Diam. In.	Circum. In.	Area Sq. In.
32	100.531	804.2	41	128.805	1320.3	50	157.080	1963.5	59	185.354	2734.0			
32⅛	100.924	810.5	41⅛	129.198	1328.3	50⅛	157.472	1973.3	59⅛	185.747	2745.6			
32¼	101.316	816.8	41¼	129.591	1336.3	50¼	157.865	1983.2	59¼	186.139	2757.2			
32⅜	101.709	823.2	41⅜	129.983	1344.3	50⅜	158.258	1993.1	59⅜	186.532	2768.8			
32½	102.102	829.5	41½	130.376	1352.7	50½	158.650	2003.0	59½	186.925	2780.5			
32⅝	102.494	835.9	41⅝	130.769	1360.8	50⅝	159.043	2012.9	59⅝	187.317	2792.2			
32¾	102.887	842.3	41¾	131.161	1369.0	50¾	159.436	2022.8	59¾	187.710	2803.9			
32⅞	103.280	848.8	41⅞	131.554	1377.2	50⅞	159.829	2032.8	59⅞	188.103	2815.7			
33	103.673	855.3	42	131.947	1385.4	51	160.221	2042.8	60	188.496	2827.7			
33⅛	104.065	861.7	42⅛	132.340	1393.7	51⅛	160.614	2052.8	60⅛	188.888	2839.2			
33¼	104.458	868.3	42¼	132.732	1402.0	51¼	161.007	2062.9	60¼	189.281	2851.0			
33⅜	104.851	874.8	42⅜	133.125	1410.3	51⅜	161.399	2073.0	60⅜	189.674	2862.9			
33½	105.243	881.4	42½	133.518	1418.6	51½	161.792	2083.1	60½	190.006	2874.8			
33⅝	105.636	888.0	42⅝	133.910	1427.0	51⅝	162.185	2093.2	60⅝	190.459	2886.6			
33¾	106.029	894.6	42¾	134.303	1435.4	51¾	162.577	2103.3	60¾	190.852	2898.6			
33⅞	106.421	901.2	42⅞	134.696	1443.8	51⅞	162.970	2113.5	60⅞	191.244	2910.5			
34	106.814	907.9	43	135.088	1452.2	52	163.363	2123.7	61	191.637	2922.5			
34⅛	107.207	914.6	43⅛	135.481	1460.7	52⅛	163.756	2133.9	61⅛	192.030	2934.5			
34¼	107.600	921.3	43¼	135.874	1469.1	52¼	164.148	2144.2	61¼	192.423	2946.5			
34⅜	107.992	928.0	43⅜	136.267	1477.6	52⅜	164.541	2154.5	61⅜	192.815	2958.5			
34½	108.385	934.8	43½	136.659	1486.2	52½	164.934	2164.8	61½	193.208	2970.6			
34⅝	108.778	941.6	43⅝	137.052	1494.7	52⅝	165.326	2175.1	61⅝	193.601	2982.7			
34¾	109.170	948.4	43¾	137.445	1503.8	52¾	165.719	2185.4	61¾	193.993	2994.8			
34⅞	109.563	955.2	43⅞	137.837	1511.9	52⅞	166.112	2195.8	61⅞	194.386	3006.9			
35	109.956	962.1	44	138.230	1520.5	53	166.504	2206.2	62	194.779	3019.1			
35⅛	110.348	969.0	44⅛	138.623	1529.2	53⅛	166.897	2216.6	62⅛	195.171	3031.3			
35¼	110.741	975.9	44¼	139.015	1537.9	53¼	167.290	2227.0	62¼	195.564	3043.5			
35⅜	111.134	982.8	44⅜	139.408	1546.6	53⅜	167.683	2237.5	62⅜	195.957	3055.7			
35½	111.527	989.8	44½	139.801	1555.3	53½	168.075	2248.0	62½	196.350	3068.0			
35⅝	111.919	996.8	44⅝	140.194	1564.0	53⅝	168.468	2258.5	62⅝	196.742	3080.3			
35¾	112.312	1003.8	44¾	140.586	1572.8	53¾	168.861	2269.1	62¾	197.135	3092.6			
35⅞	112.705	1010.8	44⅞	140.979	1581.6	53⅞	169.253	2279.6	62⅞	197.528	3104.9			
36	113.097	1017.9	45	141.372	1590.4	54	169.646	2290.2	63	197.920	3117.2			
36⅛	113.490	1025.0	45⅛	141.764	1599.3	54⅛	170.039	2300.8	63⅛	198.313	3129.6			
36¼	113.883	1032.1	45¼	142.157	1608.0	54¼	170.431	2311.5	63¼	198.706	3142.0			
36⅜	114.275	1039.2	45⅜	142.550	1617.0	54⅜	170.824	2322.1	63⅜	199.098	3154.5			
36½	114.668	1046.3	45½	142.942	1626.0	54½	171.217	2332.8	63½	199.491	3166.9			
36⅝	115.061	1053.5	45⅝	143.335	1634.9	54⅝	171.609	2343.5	63⅝	199.884	3179.4			
36¾	115.454	1060.7	45¾	143.728	1643.9	54¾	172.002	2354.3	63¾	200.277	3191.9			
36⅞	115.846	1068.0	45⅞	144.121	1652.9	54⅞	172.395	2365.0	63⅞	200.669	3204.4			
37	116.239	1075.2	46	144.513	1661.9	55	172.788	2375.8	64	201.062	3217.0			
37⅛	116.632	1082.5	46⅛	144.906	1670.9	55⅛	173.180	2386.6	64⅛	201.455	3229.6			
37¼	117.024	1089.8	46¼	145.299	1680.0	55¼	173.573	2397.5	64¼	201.847	3242.2			
37⅜	117.417	1097.1	46⅜	145.691	1689.1	55⅜	173.966	2408.3	64⅜	202.240	3254.8			
37½	117.810	1104.5	46½	146.084	1698.2	55½	174.358	2419.2	64½	202.633	3267.5			
37⅝	118.202	1111.8	46⅝	146.477	1707.4	55⅝	174.751	2430.1	64⅝	203.025	3280.1			
37¾	118.596	1119.2	46¾	146.869	1716.5	55¾	175.144	2441.1	64¾	203.418	3292.8			
37⅞	118.988	1126.7	46⅞	147.262	1725.7	55⅞	175.536	2452.0	64⅞	203.811	3305.6			
38	119.381	1134.1	47	147.655	1734.9	56	175.929	2463.0	65	204.204	3318.3			
38⅛	119.773	1141.0	47⅛	148.048	1744.2	56⅛	176.322	2474.0	65⅛	204.596	3331.1			
38¼	120.166	1149.1	47¼	148.440	1753.5	56¼	176.715	2485.0	65¼	204.989	3343.9			
38⅜	120.559	1156.6	47⅜	148.833	1762.7	56⅜	177.107	2496.1	65⅜	205.382	3356.7			
38½	120.951	1164.2	47½	149.226	1772.1	56½	177.500	2507.2	65½	205.774	3369.6			
38⅝	121.344	1171.7	47⅝	149.618	1781.4	56⅝	177.893	2518.3	65⅝	206.167	3382.4			
38¾	121.737	1179.3	47¾	150.011	1790.8	56¾	178.285	2529.4	65¾	206.560	3395.3			
38⅞	122.129	1186.9	47⅞	150.404	1800.1	56⅞	178.678	2540.6	65⅞	206.952	3408.2			
39	122.522	1194.6	48	150.796	1809.6	57	179.071	2551.8	66	207.345	3421.2			
39⅛	122.915	1202.3	48⅛	151.189	1819.0	57⅛	179.463	2563.0	66⅛	207.738	3434.2			
39¼	123.308	1210.0	48¼	151.582	1828.5	57¼	179.856	2574.2	66¼	208.131	3447.2			
39⅜	123.700	1217.7	48⅜	151.975	1837.9	57⅜	180.249	2585.4	66⅜	208.523	3460.2			
39½	124.093	1225.4	48½	152.367	1847.5	57½	180.642	2596.7	66½	208.916	3473.2			
39⅝	124.486	1233.2	48⅝	152.760	1857.0	57⅝	181.034	2608.0	66⅝	209.309	3486.3			
39¾	124.878	1241.0	48¾	153.153	1866.5	57¾	181.427	2619.4	66¾	209.701	3499.4			
39⅞	125.271	1248.8	48⅞	153.545	1876.1	57⅞	181.820	2630.7	66⅞	210.094	3512.5			
40	125.664	1256.6	49	153.938	1885.7	58	182.212	2642.1	67	210.487	3525.7			
40⅛	126.056	1264.5	49⅛	154.331	1895.4	58⅛	182.605	2653.5	67⅛	210.879	3538.8			
40¼	126.449	1272.4	49¼	154.723	1905.0	58¼	182.998	2664.9	67¼	211.272	3552.0			
40⅜	126.842	1280.3	49⅜	155.116	1914.7	58⅜	183.390	2676.4	67⅜	211.665	3565.2			
40½	127.235	1288.2	49½	155.509	1924.4	58½	183.783	2687.8	67½	212.058	3578.5			
40⅝	127.627	1296.2	49⅝	155.902	1934.2	58⅝	184.176	2699.3	67⅝	212.450	3591.7			
40¾	128.020	1304.2	49¾	156.294	1943.9	58¾	184.569	2710.9	67¾	212.843	3605.0			
40⅞	128.413	1312.2	49⅞	156.687	1953.7	58⅞	184.961	2722.4	67⅞	213.236	3618.3			

Reference Tables

CIRCUMFERENCES AND AREAS OF CIRCLES

Diam. In.	Circum. In.	Area Sq. In.	Diam. In.	Circum. In.	Area Sq. In.	Diam. In.	Circum. In.	Area Sq. In.	Diam. In.	Circum. In.	Area Sq. In.
68	213.628	3631.7	76	238.761	4536.5	84	263.894	5541.8	92	289.027	6647.6
68⅛	214.021	3645.0	76⅛	239.154	4551.4	84⅛	264.286	5558.3	92⅛	289.419	6665.7
68¼	214.414	3658.4	76¼	239.546	4566.4	84¼	264.679	5574.8	92¼	289.812	6683.8
68⅜	214.806	3671.8	76⅜	239.939	4581.3	84⅜	265.072	5591.4	92⅜	290.205	6701.9
68½	215.199	3685.3	76½	240.332	4596.3	84½	265.465	5607.9	92½	290.597	6720.1
68⅝	215.592	3698.7	76⅝	240.725	4611.4	84⅝	265.857	5624.5	92⅝	290.990	6738.2
68¾	215.984	3712.2	76¾	241.117	4626.4	84¾	266.250	5641.2	92¾	291.383	6756.4
68⅞	216.377	3725.7	76⅞	241.510	4641.5	84⅞	266.643	5657.8	92⅞	291.775	6774.7
69	216.770	3739.3	77	241.903	4656.6	85	267.035	5674.5	93	292.168	6792.9
69⅛	217.163	3752.8	77⅛	242.295	4671.8	85⅛	267.428	5691.2	93⅛	292.561	6811.2
69¼	217.555	3766.4	77¼	242.688	4686.9	85¼	267.821	5707.9	93¼	292.954	6829.5
69⅜	217.948	3780.0	77⅜	243.081	4702.1	85⅜	268.213	5724.7	93⅜	293.346	6847.8
69½	218.341	3793.7	77½	243.473	4717.3	85½	268.607	5741.5	93½	293.739	6866.1
69⅝	218.733	3807.3	77⅝	243.866	4732.5	85⅝	268.999	5758.3	93⅝	294.132	6884.5
69¾	219.126	3821.0	77¾	244.259	4747.8	85¾	269.392	5775.1	93¾	294.524	6902.9
69⅞	219.519	3834.7	77⅞	244.652	4763.1	85⅞	269.784	5791.9	93⅞	294.917	6921.3
70	219.911	3848.5	78	245.044	4778.4	86	270.177	5808.8	94	295.310	6939.8
70⅛	220.304	3862.2	78⅛	245.437	4793.7	86⅛	270.570	5825.7	94⅛	295.702	6958.2
70¼	220.697	3876.0	78¼	245.830	4809.0	86¼	270.962	5842.6	94¼	296.095	6976.7
70⅜	221.090	3889.8	78⅜	246.222	4824.4	86⅜	271.355	5859.6	94⅜	296.488	6995.3
70½	221.482	3903.6	78½	246.615	4839.8	86½	271.748	5876.5	94½	296.881	7013.8
70⅝	221.875	3917.5	78⅝	247.008	4855.2	86⅝	272.140	5893.5	94⅝	297.273	7023.4
70¾	222.268	3931.4	78¾	247.400	4870.7	86¾	272.538	5910.6	94¾	297.666	7051.0
70⅞	222.660	3945.3	78⅞	247.793	4886.2	86⅞	272.926	5927.6	94⅞	298.059	7069.6
71	223.053	3959.2	79	248.186	4901.7	87	273.319	5944.7	95	289.451	7088.2
71⅛	223.446	3973.1	79⅛	248.579	4917.2	87⅛	273.711	5961.8	95⅛	298.844	7106.9
71¼	223.838	3987.1	79¼	248.971	4932.7	87¼	274.104	5978.9	95¼	299.237	7125.6
71⅜	224.231	4001.1	79⅜	249.364	4948.3	87⅜	274.497	5996.0	95⅜	299.629	7144.3
71½	224.624	4015.2	79½	249.757	4963.9	87½	274.889	6013.2	95½	300.022	7163.0
71⅝	225.017	4029.2	79⅝	250.149	4979.5	87⅝	275.282	6030.4	95⅝	300.415	7181.8
71¾	225.409	4043.3	79¾	250.542	4995.2	87¾	275.675	6047.6	95¾	300.807	7200.6
71⅞	225.802	4057.4	79⅞	250.935	5010.9	87⅞	276.067	6064.9	95⅞	301.200	7219.4
72	226.195	4071.5	80	251.327	5026.5	88	276.460	6082.1	96	301.593	7238.2
72⅛	226.587	4085.7	80⅛	251.720	5042.3	88⅛	276.853	6099.4	96⅛	301.986	7257.1
72¼	226.980	4099.8	80¼	252.113	5058.0	88¼	277.246	6116.7	96¼	302.378	7276.0
72⅜	227.373	4114.0	80⅜	252.506	5073.8	88⅜	277.638	6134.1	96⅜	302.771	7294.9
72½	227.765	4128.2	80½	252.898	5089.6	88½	278.031	6151.4	96½	303.164	7313.8
72⅝	228.158	4142.5	80⅝	253.291	5105.4	88⅝	278.424	6168.8	96⅝	303.556	7332.8
72¾	228.551	4156.8	80¾	253.684	5121.2	88¾	278.816	6186.2	96¾	303.949	7351.8
72⅞	228.944	4171.1	80⅞	254.076	5137.1	88⅞	279.209	6203.7	96⅞	304.342	7370.8
73	229.336	4185.4	81	254.469	5153.0	89	279.602	6221.1	97	304.734	7389.8
73⅛	229.729	4199.7	81⅛	254.862	5168.9	89⅛	279.994	6238.6	97⅛	305.127	7408.9
73¼	230.122	4214.1	81¼	255.254	5184.9	89¼	280.387	6256.1	97¼	305.520	7428.0
73⅜	230.514	4228.5	81⅜	255.647	5200.8	89⅜	280.780	6273.7	97⅜	305.913	7447.1
73½	230.907	4242.9	81½	256.040	5216.8	89½	281.173	6291.2	97½	306.305	7466.2
73⅝	231.300	4257.4	81⅝	256.433	5232.8	89⅝	281.565	6308.8	97⅝	306.698	7485.3
73¾	231.692	4271.8	81¾	256.825	5248.9	89¾	281.958	6326.4	97¾	307.091	7504.5
73⅞	232.085	4286.3	81⅞	257.218	5264.9	89⅞	282.351	6344.1	97⅞	307.483	7523.7
74	232.478	4300.8	82	257.611	5281.0	90	282.743	6361.7	98	307.876	7543.0
74⅛	232.871	4315.4	82⅛	258.003	5297.1	90⅛	283.136	6379.4	98⅛	308.269	7562.2
74¼	233.263	4329.9	82¼	258.396	5313.3	90¼	283.529	6397.1	98¼	308.661	7581.5
74⅜	233.656	4344.5	82⅜	258.789	5329.4	90⅜	283.921	6414.9	98⅜	309.054	7600.8
74½	234.049	4359.2	82½	259.181	5345.6	90½	284.314	6432.6	98½	309.447	7620.1
74⅝	234.441	4373.8	82⅝	259.574	5361.8	90⅝	284.707	6450.4	98⅝	309.840	7639.5
74¾	234.834	4388.5	82¾	259.967	5378.1	90¾	285.100	6468.2	98¾	310.232	7658.9
74⅞	235.227	4403.1	82⅞	260.359	5394.3	90⅞	285.492	6486.0	98⅞	310.625	7678.3
75	235.619	4417.9	83	260.752	5410.6	91	285.885	6503.9	99	311.018	7697.7
75⅛	236.012	4432.6	83⅛	261.145	5426.9	91⅛	286.278	6521.8	99⅛	311.410	7717.1
75¼	236.405	4447.4	83¼	261.538	5443.3	91¼	286.670	6539.7	99¼	311.803	7736.6
75⅜	236.798	4462.2	83⅜	261.930	5459.6	91⅜	287.063	6555.6	99⅜	312.196	7756.1
75½	237.190	4477.0	83½	262.323	5476.0	91½	287.456	6575.5	99½	312.588	7775.6
75⅝	237.583	4491.8	83⅝	262.716	5492.4	91⅝	287.848	6593.5	99⅝	312.981	7795.2
75¾	237.976	4506.7	83¾	263.108	5508.8	91¾	288.241	6611.5	99¾	313.374	7814.8
75⅞	238.368	4521.5	83⅞	263.501	5525.3	91⅞	288.634	6629.6	99⅞	313.767	7834.4
									100	314.159	7854.0

THICKNESSES AND WEIGHTS OF METALS

Carbon Steel (black) • Galvanized • Stainless

GAGE	CARBON STEEL (black) Mfr's. Std. Gage	Lbs. per Sq. Ft. Decimal Inches	Fraction	GALVANIZED Mfr's. Thickness for Steel + .0037"	Weight Per Square Foot, Pounds	U. S. Standard Approximate Decimal Parts of an Inch Thickness	STAINLESS Lbs. per Sq. Ft. Chrome	Chrome Nickel
10	.1345	5.625	5⅝	.1382	5.781	.140	5.793	5.906
12	.1046	4.375	4⅜	.1084	4.531	.109	4.506	4.593
14	.0747	3.125	3⅛	.0785	3.281	.078	3.218	3.281
16	.0598	2.500	2½	.0635	2.656	.062	2.575	2.625
18	.0478	2.000	2	.0516	2.156	.050	2.060	2.100
20	.0359	1.500	1½	.0396	1.656	.037	1.545	1.575
22	.0299	1.250	1¼	.0336	1.406	.031	1.287	1.312
24	.0239	1.000	1	.0276	1.156	.025	1.030	1.050
26				.0217	.906	.018	.772	.787
28				.0187	.781	.015	.643	.656
30				.0157	.656	.012	.515	.525

349

WEIGHT OF GALVANIZED IRON

FULL SHEETS

| Gage No. | WEIGHT IN POUNDS | | | | |
| | 36" Width | | 48" Width | | Per Square Foot |
	Length 96"	Length 120"	Length 96"	Length 120"	
10	138.75	173.44	185.00	231.22	5.781
12	108.75	135.94	145.00	181.24	4.531
14	78.75	98.44	105.00	131.24	3.281
16	63.75	79.69	85.00	106.24	2.656
18	51.75	64.69	69.00	86.24	2.156
20	39.75	49.69	53.00	66.24	1.656
22	33.75	42.19	45.00	56.25	1.406
24	27.75	34.69	37.00	46.24	1.156
26	21.75	27.19	29.00	36.24	.906
28	18.75	23.44	25.00	27.30	.781
30	15.75	19.70			.656

Equivalent Thicknesses
DUCT

| Galvanized | | | Aluminum | |
#	Mfr. Ga.	SIZE IN INCHES	#	B/S Ga.
26	.018	0 — 12	24	.020
24	.024	13 — 30	22	.025
22	.030	31 — 60	20	.032
20	.036	61 — 90	18	.040
18	.048	91 — up	16	.050

index

index

index